Micro- and Nanofabrication
for Beginners

Micro- and Nanofabrication for Beginners

Eiichi Kondoh

JENNY STANFORD
PUBLISHING

Published by

Jenny Stanford Publishing Pte. Ltd.
Level 34, Centennial Tower
3 Temasek Avenue
Singapore 039190

Email: editorial@jennystanford.com
Web: www.jennystanford.com

British Library Cataloguing-in-Publication Data
A catalogue record for this book is available from the British Library.

Micro- and Nanofabrication for Beginners

Copyright © 2021 Jenny Stanford Publishing Pte. Ltd.

This book is based on the author's publication from Kyoritsu Shuppan Co., Ltd., Japan, and is for distribution only outside Japan.

ISBN 978-981-4877-09-1 (Hardcover)
ISBN 978-1-003-11993-7 (eBook)

Contents

Preface

Atoms and molecules are the smallest units of substances. To understand the principles of micro- and nanofabrication, it is important to go back to the scales as small as molecules and atoms of physical and chemical phenomena. The purpose of this book is to describe the physical and chemical principles of micro- and nanofabrication based on the concept of using gas molecules as a "processing tool." The book will be useful for undergraduates and graduates of chemistry, materials science, mechanical engineering, and electronic engineering, as well as young researchers and engineers who have just become a part of this field.

Numerous novel technologies have been developed and researches have been done for micro- and nanoprocessing. From a practical point of view, however, thin-film dry processing is the sole technology that has been used in manufacturing so far. Thin-film dry processing is essential in fabricating modern advanced devices, such as integrated circuits, storage devices, flat displays, and sensors, that involve a wide range of study areas, including mechanical engineering, precision engineering, electronics, chemistry, physics, and materials manufacturing. Thin-film processing finds wide applications and is the basic technology to realize novel devices such as micromachines/microelectromechanical systems (MEMS), and medical chips. It is obvious that the science and techniques of thin-film processing are noteworthy subjects to be learned over time.

Thin-film processing is a composite technology that consists of different elementary processes and techniques. A different method, based on various physical and chemical principles, is employed in each process. To compile a series of chapters to describe each process can be quite boring and even confusing for readers, therefore, it is more crucial to learn and understand the core points of physics and chemistry of thin-film processing. For these reasons, this book focuses on the basics of thin-film processing from the point of view of "using gases as machinery tools."

Most of the existing textbooks or specialized books have been written on the premise of fabrication of electronic devices, which hinders sound understanding of beginners who do not have a solid background of electronic devices. On the other hand, electronic engineers do not always have a solid background of physicochemistry or materials science that are essential to learn thin-film processing. To understand this book, an educational background of electronic devices or electromagnetics is not necessary, and anyone with a knowledge of physical basics of high school or university fresher year can find it useful. It is often important not just to learn about the models or equations but also to understand their core concepts. In this book, physical models and their equations have been carefully described. In addition, important equations and concepts have been emphasized using text boxes. I hope this book will help readers to understand the much extended subjects in advanced and specialized books and literature.

Eiichi Kondoh
January 2021

Chapter 1

Introduction

1.1 Technologies That Underline the Information Society

In 1949, after World War II, the first all-electronic computer ENIAC (Electronic Numerical Integrator and Computer) was invented at Pennsylvania University in the United States. ENIAC was a huge machine equipped with about 18,000 vacuum tubes and spanned approximately 80 m and was able to perform 5,000 operations per second—it was really a revolutionary computing apparatus of the era.

In 1996, in commemoration of the 50th anniversary of the birth of ENIAC, Pennsylvania University fabricated a computer chip that functioned exactly as the ENIAC. However, it was really small, only 7.44 mm × 5.29 mm in size and manufactured by the leading-edge 0.5 μm triple-layer metal interconnect CMOS technology. Despite its smallness, the chip had 17,000 transistors and, amazingly, performed 10 billion operations per second!

What brought so much difference in the performance between these two computers that have almost the same architecture? It is microfabrication that had progressed. As dimensions of electronic

Micro- and Nanofabrication for Beginners
Eiichi Kondoh
Copyright © 2021 Jenny Stanford Publishing Pte. Ltd.
ISBN 978-981-4877-09-1 (Hardcover), 978-1-003-11993-7 (eBook)
www.jennystanford.com

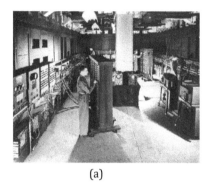
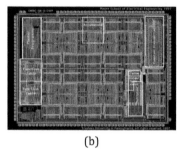

(a)   (b)

Figure 1.1 The world's first fully-electronic computer ENIAC (a) and an ENIAC-chip (b) (http://www.ee.upenn.edu/jan/eniacproj.html).

devices shrink, the devices function faster with lower electric power (Fig. 1.1).

Microfabrication allows miniaturization as well as lightweighting of electronic equipments. If the components are smaller in size, more of them and electronic devices can be embedded in a same area or volume, which inturn would help in delivering higher performance. High growth can be seen in information technology due to micorfabrication technology.

1.2 World of Micro and Nano

The term "micro" is an adjective or prefix meaning "extremely small." It is a metric prefix that indicates 10^{-6}. For instance, 1 μm (one micrometer or one micron) indicates 10^{-6} m. "Microfabrication" cannot be defined very clearly and its usage may differ from person to person or community to community. It can commonly be understood in two ways: to fabricate extremely small components regardless of their absolute dimensions, or to fabricate components approximately of 1 μm size or smaller than that.

Recently the term "nanotechnology" has become very popular. The origin of "nano" is a Greek word ναvōς. The metric prefix "nano" indicates 10^{-9} and 1 nm (one nanometer) is 10^{-9} m. "Nanofabrication" is a fabrication technology used in nanotechnology. Nanofabrication is employed for fabricating components

Table 1.1 Unit prefixes

yotta	Y	10^{24}	deci	d	10^{-1}
zetta	Z	10^{21}	centi	c	10^{-2}
exa	E	10^{18}	milli	m	10^{-3}
peta	P	10^{15}	micro	μ	10^{-6}
tera	T	10^{12}	nano	n	10^{-9}
giga	G	10^{9}	pico	p	10^{-12}
mega	M	10^{6}	femto	f	10^{-15}
kilo	k	10^{3}	atto	a	10^{-18}
hecto	h	10^{2}	zepto	z	10^{-21}
deca	da	10	yocto	y	10^{-24}

smaller than the ones fabricated by microtechnology, generally smaller than 1 μm, and the definition of nanofabrication is also not very clear.

Can you realize how small is the nanoworld? Let us see Fig. 1.2. This figure lists sizes of various objects having different dimensions on the same scale called the logarithmic scale. The size decuple by each graduation to the left and becomes 1/10 times to the right. Height of a human being is "approximately" 1 m and is placed at the center of the scale. The difference in scale between a human being and the Tokyo tower standing at 333 m—a communications and observation tower located in Tokyo—is two graduations or 10^2 times. An object two-graduations smaller than a human being is an insect. That is, an insect views a human being as we view Tokyo tower.

The diameter of a human hair is approximately 200–300 μm. *E. coli* cells are a few μm long, and a DNA has a diameter of 2 nm. The minimum dimension of components of integrated circuits is 10 nm (and below 1 nm in thickness). This size is smaller than or as small as that of viruses. It is about 10^{-8} times smaller than human beings. Then to what are human beings 1×10^{-8} smaller? Human beings have a scale of 1 m and an object 8 orders of magnitude larger has a scale of 10,000 km. It is as large as the Earth. In micro-/ nanotechnologies, components of 10 nm are fabricated.[1] One can

---

[1] Sizes of viruses are 10–100 nm. Human beings appear something like viruses to the Earth. Are we viruses of the Earth?

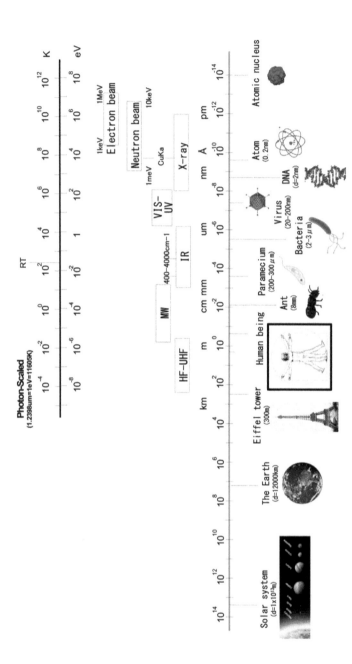

Figure 1.2 Objects having various sizes placed on the same scale. One-graduation difference is identical to one digit (10 times).

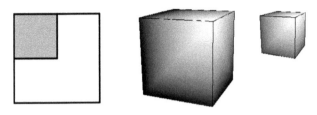

Figure 1.3 By downsizing a component by half, four times more components can be embedded in same area and eight times more in a same space.

imagine that placing a nanocomponent in a device is same as a human-sized object being placed somewhere on Earth, e.g., a stool in a room.

By downsizing the components by 1/2 times, 4 ($= 2^2$) times more components can be embedded in a given same area, and 8 times more in a given volume (Fig. 1.3). Downsizing gives us the benefit of more available space. The size of atoms is a few Å^2 and 10^{-11} times smaller than human beings. The diameter of Mercury's orbit is 10^{10} times larger than that of atoms. Human beings have explored and expanded spaces from local regions to continents and to oceans, and they are very likely to expand it to space in the near future. More spaces are also opening up as micro- and nanospaces, if we travel in the reverse direction on the scale.

1.3 Contents and Construction of This Book

Atoms and molecules are the smallest units of matter. In order to understand the fundamentals of micro- and nanofabrication, it is necessary to decipher various fabrication processes at the atomic or molecular level. This book explains the fundamentals of micro- and nanofabrication using the concept of gaseous molecules as a "tool."

In Chapter 2, we study the behavior of gases. Chapter 4 describes the fundamentals of plasmas as a means of controlling the motion and chemical behavior of gases. Thin films are raw materials of micro- and nanocomponents, and how to develop a thin film from gases is described in Chapter 4. Following that, in

---

[2]$1 \text{ Å} = 0.1 \text{ nm} = 10^{-9} \text{ m}.$

Chapter 5, we study about the process of developing thin films along with their materials science. Thin films can be patterned into micro- and nanocomponents. Material removal—called etching—by gases is described in Chapter 6 along with solution-based etching techniques. Photolithography (Chapter 7) is a series of techniques that transfers a pattern using an optical method.

Hold this book and go to the nanoworld. Bon voyage!

Chapter 2

Vacuum and Gas Kinetics

2.1 Introduction

What kind of tools are needed for microscale and nanoscale fabrication? Would one be surprised to know that gases are used as the tools?

The smallest units of matter are atoms and molecules. To realize very precise fabrication, the tools needed must have preciseness as high as atoms and molecules. In micro- and nanofabrication, materials are subtracted and added by irradiating rarefied gases to a work part. This gas-based process is called a vacuum process or dry process. Before we begin to learn the fundamentals of micro- and nanoprocesses and the materials, we will learn the physics of vacuum and vacuum apparatuses.

2.2 Vacuum and Equation of State

2.2.1 Vacuum

Vacuum is, intrinsically, a vacant space of matter and is conceptualistic. In science and engineering, vacuum is a space filled with

Micro- and Nanofabrication for Beginners
Eiichi Kondoh
Copyright © 2021 Jenny Stanford Publishing Pte. Ltd.
ISBN 978-981-4877-09-1 (Hardcover), 978-1-003-11993-7 (eBook)
www.jennystanford.com

Table 2.1 Vacuum ranges (Pa)

Low vacuum	Atmospheric pressure to 10^2
Middle vacuum	10^2–10^{-1}
High vacuum (HV)	10^{-1}–10^{-6}
Ultra-high vacuum (UHV)	10^{-6}–10^{-9}
Extremely high vacuum (XHV)	$<10^{-9}$

rarefied gases at pressure lower than atmospheric pressure. The term vacuum is used to indicate a very wide range of pressure from sub-atmospheric pressure to reduced pressure, and to a space very close to real (conceptualistic) vacuum. Vacuum does not mean pressure.

Even cosmic space is filled with very dilute gases, and therefore, it is a vacuum in the sense of science and engineering. The pressure at 1000 km above the ground is 10^{-9} Pascal (Pa). Such vacuum is very difficult to realize industrially or even experimentally. The orbit altitude of space shuttles and the International Space Station is 400 km and the vacuum pressure there is approximately 10^{-6} Pa. The pressure of the air of Mars is 6 hPa.

The level of vacuum is expressed by the pressure of gases. A vacuum filled with a dilute gas has a "low pressure" and vise versa. Commonly, a "good vacuum" means a low-pressure environment because pressure reduction is always a challenge in vacuum technology. Table 2.1 lists categories of vacuum ranges.

"Pascal" is the SI unit (Le Système International d'Unités)[1] that measures pressure. Pressure is the force per unit area and Pa—identical to N/m^2—is used to indicate values of pressure (or stress) even for pressures higher than atmospheric pressure. In vacuum science and technology, various units have been used traditionally and Table B.2 lists some of those. Most of the recent instruments use only the SI units and may optionally display other old units. Old instruments do not even have an SI scale. Table B.3 is a conversion

---

[1]The International System of Units is the metric system internationally agreed upon and is used in most countries, both in sciences and everyday living. The SI units are coherently constructed based on the seven basic units (s, m, kg, A, K, mol, cd). Other units are derived from these units.

table of units for vacuum or pressure. Unusual conversion factors are intentionally left blank.

Advantages of vacuum are as follows:

(i) It is easy to control the collision between gas molecules and between gas molecules and surface because the gas is dilute.
(ii) It is easy to realize a nonequilibrium environment.
(iii) It is easy to control the level of impurities, such as water.
(iv) It is easy to achieve a precise control of the gaseous composition or density.
(v) It has a uniform atmosphere because the gas expands rapidly.
(vi) It acts as a thermal insulator because it neither convects or conducts heat.
(vii) It can be used as a force, such as in a vacuum chuck, vacuum cleaner, etc.

Vacuum processing or dry processing became popular for reasons (i)–(iv) and are now widely used for thin film deposition, etching, surface cleaning, and surface modification.

2.2.2 Ideal Gases and Behavior of Gases

N mol of **ideal gas** at a pressure of p and temperature of T has a volume V. This relation is called the **equation of state** or perfect gas equation,

$$pV = NRT \qquad (2.1)$$

where R is the gas constant. $R = 8.314510$ (J/mol/K) for the unit system of Pa for p, K for T, and m^3 for V. $R = 8.205 \times 10^{-2}$ ($\ell \cdot$ atm/mol/K) for atm, K, ℓ (liter). Basic physical constants for gas kinetics are listed in Table B.1 of Appendix B.

An ideal gas consists of molecules[2] between which attractive (or repulsive) interactions are negligible. We can ignore interactive

---

[2]A molecule is the smallest entity of a substance that retains the characteristics of that substance. Molecules of rare gases, such as Ar and He, are single atoms. In physics of vacuum and gas kinetic theory, a molecule does not always mean a chemical molecule. It can also mean a particle that makes up a gas. We deal with ions and electrons as molecules.

forces when

$$\begin{cases} \text{Molecular interactions are very weak, such as in rare gases.} \\ \text{The gas is sufficiently rarefied.} \end{cases} \quad (2.2)$$

Strictly speaking, no real gas satisfies these conditions and thus there is no ideal gas. However, feel safe, in most cases of vacuum, a gas can be treated as an ideal gas.

The following equation tells the number of molecules in a volume at any given temperature and pressure. For instance, the number of molecules confined in 1 m^3 under the standard state condition can be obtained by using the fact that 1 mol $\equiv N_A$,

$$\frac{N}{V}N_A = \frac{p}{RT} = \frac{1.013 \times 10^5}{8.3145 \times 273} \times 6.02 \times 10^{23} = 2.69 \times 10^{25} \quad (2.3)$$

That is, 1 ml contains 2.69×10^{19} molecules of gases. This shows that numerous molecules exist in a very small volume.

The number of molecules contained in a unit volume is known as number density.

When pressure is reduced under constant temperature, the number density decreases, and when number density is kept constant, the temperature decreases.

Rewriting Eq. (2.1) as

$$p = nkT \quad\quad\quad\quad (2.4)$$

where $n = NN_A/V$. The unit of number density is m^{-3}.

The ideal gas law states that pressure, temperature, and density are identical quantities. Hereafter, we will call the above equation "ideal gas law at molecular level," in this book because this equation is a rewrite of Eq. (2.1) with respect to gas molecules. This equation shows a direct connection between pressure and temperature.

2.3 Gas Pressure and Internal Energy

Even a miniscule volume contains numerous gas molecules that travel at a breakneck speed. Their motion is translatory, and

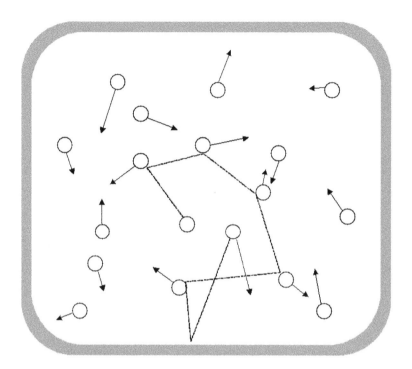

Figure 2.1   Motion of gas molecules filled in a volume.

therefore, its direction changes very frequently on collision with the vacuum chamber wall or other molecules (Fig. 2.1). As a result, it becomes a perfectly random motion. Therefore the motion is analyzed statistically.

2.3.1 Gas Pressure and Speed of Gas Molecules

Gas molecules move freely and randomly in a container. Pressure is the force that the molecules exert on colliding with the wall of the container.

Let us assume that a gas molecule is translating in space with speed $\mathbf{v}$. The components of $\mathbf{v}$ are v_x, v_y, v_z. The scalar v of $\mathbf{v}$ is

$$v^2 = v_x{}^2 + v_y{}^2 + v_z{}^2 \qquad (2.5)$$

We will first examine only one direction, x. The momentum of the molecule is mv_x where m is the mass of the molecule. When

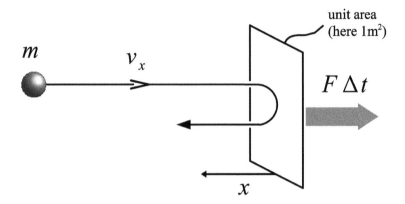

Figure 2.2 Motion of gas molecules filled in a volume.

the molecule collides elastically and vertically to a wall, the speed of the reflected molecule is $-v_x$, and the change of momentum is $2mv_x$ (speed components y, z do not change).[3]

The number of molecules colliding to the wall in an infinitesimal time interval Δt was contained in a volume of $v_x \Delta t$ ($\times$ area) and is therefore $nv_x \Delta t$. In average, the number of molecules travelling in the $+$ direction and in the $-$ direction is the same, and therefore, half of the above molecules $nv_x \Delta t/2$ are moving toward the wall. The total change of the momentum due to the collision is (Fig. 2.2)

$$2mv_x \frac{nv_x \Delta t}{2} = mnv_x^2 \Delta t \tag{2.6}$$

The reaction force is the change of the momentum $\left(= \dfrac{dp}{dt} = m\dfrac{dv}{dt} \right)$, and the total reaction force, or pressure, is

$$p = nmv_x^2 \tag{2.7}$$

---

[3] The mass of a molecule is

$$m = \frac{M_w}{N_A}$$

where M_w is the molar mass (kg/mol) of the molecule. The molar mass is a physical constant and has a value identical to the molecular weight but its unit is gram. It should be noted that whenever values of m or M_w are substituted in formulas in this book, the unit kg should be used as the dimension of mass. For instance, the mass of a single N_2 molecule (as its $M = 28$) is $m = 28 \times 10^{-3}/6.02 \times 10^{23} = 4.65 \times 10^{-26}$ kg.

In reality, all molecules travel with different speeds, and the distribution range of their speeds are given in Section 2.5. We then replace v_x^2 with $\overline{v_x^2}$,

$$p = nm\overline{v_x^2} \tag{2.8}$$

Now we can extend this discussion to three-dimensional space. As all molecules are moving perfectly randomly, the overall motion in x, y, z directions are independent and identical. From vector mathematics, we know $\overline{v_x^2} = \overline{v_y^2} = \overline{v_z^2}$ and $\overline{v^2} = \overline{v_x^2} + \overline{v_y^2} + \overline{v_z^2}$. Therefore, we obtain

$$p = \frac{nm\overline{v^2}}{3} \tag{2.9}$$

which is an equation that shows the relationship between pressure and kinetic energy.

2.3.2 Internal Energy of a Gas

The kinetic energy of one gas molecule moving at a speed v is $mv^2/2$. The number of molecules in 1 m^3 is n. The total kinetic energy is

$$E = \frac{1}{2}nm\overline{v^2} \tag{2.10}$$

The internal energy of a gas is the sum of the kinetic energy and the potential energy of the gas molecules. The potential energy stems from the interactive forces between molecules and is negligible for an ideal gas. Therefore, E is identical to the **internal energy** or thermal energy of a gas.

From the ideal gas law at molecular level Eq. (2.4) and the relationship between pressure and kinetic energy Eq. (2.9), we obtain

$$p = nkT = \frac{nm\overline{v^2}}{3} = \frac{2}{3}E \tag{2.11}$$

and the energy per molecule is

$$\frac{1}{2}m\overline{v^2} = \frac{3}{2}kT \tag{2.12}$$

This equation expresses **the relationship between the kinetic energy and the temperature of a molecule**. It is obvious that temperature cannot have a negative value because $E \geq 0$.

So far we have assumed that gas molecules behave as a point object. The degree of freedom[4] is 3. The mean kinetic energy of a molecule is distributed equally to each freedom, and the energy per freedom is

$$\frac{kT}{2} \tag{2.13}$$

(equipartition theorem). The degree of freedom of diatomic molecules, such as H_2 and N_2, is 5, and the coefficient of Eq. (2.12) is 5/2.

The source of energy of the motion of molecules is thermal energy. In an ideal gas, heat or internal energy is identical to the kinetic energy of molecules. When a vessel in which a gas is enclosed is heated, the heat transfers from the wall of the vessel to the gas molecules inside it and the temperature and/or pressure of the gas increases. Molecules colliding with the wall gain the heat energy and when they collide with each molecules they distribute the thermal energy from the wall to all gas molecules.

2.4 Total Pressure and Partial Pressure

For a mixed gas consisting of different gaseous species, the **total pressure** p_T is the sum of pressures of each species p_1, p_2, $p_3 \cdots$. The pressure of each gaseous species is called **partial pressure**. If you extract only one gas component and enclose it in another vessel, the total pressure of the vessel is identical to the partial pressure of

---

[4]Degree of freedom is the minimum number of dimensions of a space system that describes the motion of an object in that system. The number of dimensions that defines the position of a point object is three, and thus the degree of freedom is 3. More information is needed to define a complex object. To define the motion of a line segment, its direction and rotation are introduced as other independent dimensions, and therefore, its degree of freedom is 5. The motion of a simple spherical gas molecule is translational and its degree of freedom is 3.

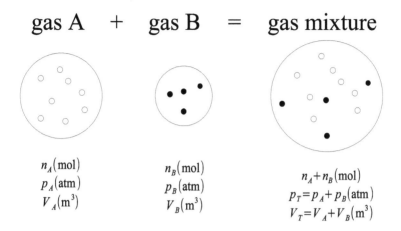

Figure 2.3 Total pressure and partial pressures.

the original mixture.

$$p_T = \sum_i p_i = \frac{RT}{V} \sum_i n_i = kT \sum_i n_i \qquad (2.14)$$

Therefore, the partial pressure is proportional to the concentration (molar concentration or number density) of the gas component.

The concentration of oxygen in the atmosphere is 21%, and its partial pressure in the atmosphere is $p_{O_2} = 0.21$ [atm] $= 2.1 \times 10^5$ [Pa].

Partial pressure does not simply show the ratio of gas components. The rate of chemical reactions of the gases involved is proportional to the partial pressures of the gases. For instance, rate of oxidation becomes smaller as p_{O_2} decreases.

Let us assume that we have a completely exhausted vacuum chamber and gaseous O_2 is added to it at a pressure of 100 Pa. The total pressure and the O_2 partial pressure is the same, $p_{O_2} = 100$ [Pa]. Next, you enclose the atmosphere in a different chamber and blow N_2 into the chamber. The oxygen content or p_{O_2} decreases gradually and will reach 100 Pa. In view of p_{O_2}, these two atmosphere are identical.[5]

---

[5] Coexisting gaseous species can influence on or compete with chemical processes.

2.5 Distribution Law of Gas Velocity

2.5.1 Maxwell–Boltzmann Gas Velocity Distribution

Gas molecules move randomly and freely. Even if all the molecules have the same velocity initially, or kinetic energy, they collide with each other extremely frequently and molecules having very large kinetic energy and almost zero energy will appear. Gas velocity v shows a wide spread.

Motion of individual molecules is probabilistic, and a statistical approach is needed to treat problems of the behavior of the total gas molecules. In gas kinetic statistics, **distribution function of molecular speeds** shows the distribution of the probability that molecules at a particular speed will appear.[6]

Maxwell–Boltzmann distribution is a distribution function of molecular speeds.

$$f(v) = 4\pi v^2 \left(\frac{m}{2\pi kT}\right)^{\frac{3}{2}} \exp\left(-\frac{mv^2}{2kT}\right) \qquad (2.15)$$

Figure 2.4 shows profiles of this function. The maxima appears at a speed of v_m. No molecules have a speed of 0 or ∞.

Distribution profile depends on gas temperature T and mass of a gas molecule m, or molecular weight M_w. The profile becomes sharper and narrower when temperature and/or molecular weight is low. More precisely, from Eq. (2.15), we understand that $f(v)$ is a function of m/T, indicating that low temperature and higher molecular weight are identical. This is quite obvious because internal energy of a gas is identical to the total kinetic energy of the gas molecules [Eq. (2.9)].

2.5.1.1 Mean speeds and most probable speed

Several "mean speeds" are defined in gas kinetics as can be seen in Table 2.2.

---

[6]In mathematics of statistics, this function is called probability density function.

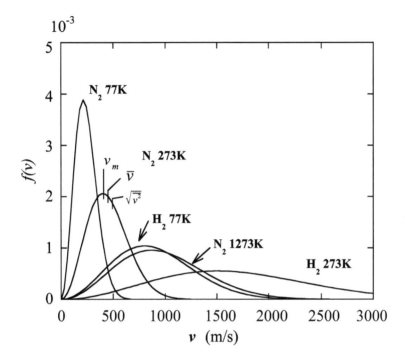

Figure 2.4 Maxwell–Boltzmann distribution.

Table 2.2 Mean speeds

$\bar{v}$	Mean speed
$\overline{v^2}$	Mean square speed
$\bar{v}^2$	Squared mean speed $(\overline{v^2} \neq \bar{v}^2)$

$f(v)$ takes 0 at $v = 0$ and $v = \infty$. The maxima exists in-between these values and is called the **most probable speed** of gas molecule. The most probable speed is obtained by setting 0 to the derivative of the Maxwell–Boltzmann function.

$$v_m = \sqrt{\frac{2kT}{m}} \qquad (2.16)$$

Square speeds are important as kinetic energy of a gas molecule is proportional to them. Mean square speed is obtained from Eq. (2.12), however, a usual speed expression is more convenient.

Table 2.3 Mean speeds of molecules at 273 K (m/s)

	M_w	$\bar{v}$	$\sqrt{\bar{v^2}}$	v_m
H_2	2	1.70×10^3	1.85×10^3	1.51×10^3
N_2	28	454	493	403
O_2	32	425	461	377
H_2O	18	567	615	502
Ar	40	380	413	337
Relative ratio		1.13	1.22	1

For this reason, we usually use **root-mean-square speed,**

$$\sqrt{\bar{v^2}} = \sqrt{\frac{3kT}{m}} \qquad (2.17)$$

The following relation is obvious from Eq. (2.16).[7]

$$\sqrt{\bar{v^2}} = \sqrt{\frac{3}{2}} v_m \qquad (2.18)$$

It should be noted that mean-square speed $\bar{v^2}$ is a mean of square of v and is not equal to square of average speed $\bar{v}^2$.

The mean speed is given by

$$\bar{v} = \int_0^\infty v f(v) dv = \sqrt{\frac{8kT}{\pi m}} = \frac{2}{\sqrt{\pi}} v_m \qquad (2.19)$$

The mean speed is almost equal to the most probable speed. Either the mean speed or the most probable speed is independent of gas pressure. If the temperature is same, these speeds are same in vacuum as well as in the atmosphere.

Table 2.3 lists mean speeds at 273 K obtained by using Eq. (2.18) and Eq. (2.19).

2.5.1.2 Meaning of distribution function

What does the distribution function $f(v)$ tells us? How many molecules are flying at the speed v? Fig. 2.4 Let us calculate the

---

[7]It is practical to use molecular weight M_w when obtaining $\sqrt{\bar{v^2}}$. As k is a constant, only M_w, T are independent variables in Eq. (2.17). Now we obtain a simplified expression $158\sqrt{T/M_w}$ (m/s). The root-mean-square speed of N_2 molecules at 273 K is $158\sqrt{299/28} = 493$ (m/s). It should be noted that this equation always gives a value with a unit of m/s.

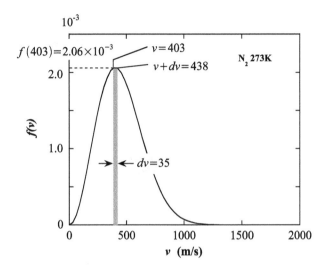

Figure 2.5 Illustrative drawing showing how to calculate a number of molecules travelling at v from a distribution function $f(v)$.

number of N_2 molecules travelling at a speed of v at 298 K by using the speed distribution shown in Section 2.5.1.1.

First, calculate an area of a bar bounded by $f(v)$ and infinitesimal interval $(v, v + dv)$. A representative value of this interval is v and we here use most probable speed, $v_m = 403$. We here assume an interval $dv = 35$. (The smaller the dv, the better accuracy we can expect.) The value of f at the representative value of $v = 403$ is $f(403) = 2.06 \times 10^{-3}$, and the area is

$$dvf(v) = 35 \times 2.06 \times 10^{-3} = 7.21 \times 10^{-2} \qquad (2.20)$$

See Eq. (2.15) and notice that the unit of v^2 is m^2/s^2 and the unit of $m/2kT$ is $kg/J \equiv s^2/m^2$. The unit of $f(v)$ is the inverse of speed, s/m, and therefore, $dvf(v)$ is nondimensional.

This nondimensional value shows the appearance probability of molecules within the interval of $(v, v + dv)$. That is, 7.21% of total gas molecules have a speed between 403 m/s and 438 m/s. At atmospheric pressure, molecular density is $n = 2.46 \times 10^{25}$ m^{-3}, and thus the number of molecules that have a speed between 403 m/s and 438 m/s is,[8]

$$ndvf(v) = 2.46 \times 10^{25} \times 7.21 \times 10^{-2} = 1.78 \times 10^{24} \text{ per 1 m}^3.$$

$$(2.21)$$

As understood from the above, the distribution function $f(v)$ is the existence probability of v within an interval $(v, v + dv)$ divided by dv. This quantity is a weight or density.[9]

Reconsider Eq. (2.8) in which pressure and density were related, we had simply replaced $v_x{}^2$ by $\overline{v_x{}^2}$. Now since we understand the meaning of the probability function, we can calculate the mean square speed of gas molecules.

The number of molecules that collide with a wall is $v_x dn(v_x)\Delta t$ per unit area for a time interval Δt, and the change of the kinetic momentum is $mv_x{}^2 dn(v_x)\Delta t$ from Eq. (2.6). The total reaction force is expressed by the following Eq. (2.7).

$$p = \int_{v_x=0}^{v_x=\infty} 2mv_x{}^2 dn(v_x)$$

$$= mn \int_0^\infty v_x{}^2 f(v_x) dv_x$$

$$= mn\overline{v_x{}^2} \qquad (2.22)$$

By substituting the one-dimensional distribution function and calculating the integral, we obtain $m\overline{v_x^2} = kT$ [see Eq. (2.13)].

2.5.2 Number of Molecules Passing Through a Unit Area

2.5.2.1 Flux of incident molecules

Molecules travel freely and independently in a vacuum keep colliding with the wall of the vacuum chamber. When viewed from the wall, one can see molecules coming from various directions and hitting it.

Now imagine a cube having a unit volume and containing n molecules. If the cube collides with the wall, then n molecules hit the wall. If the cube has an average velocity of $\bar{v}$, the wall is hit by n molecules within a time span of $1/v$ (because the length of one side

---

[8]$dn(v) = nf(v)dv = n(v)dv.$
[9]In mathematical statistics, this function is called density function.

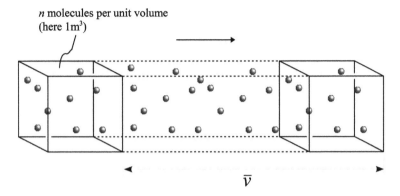

n molecules per unit volume
(here 1m³)

$\overline{v}$

Figure 2.6 One-dimensional model of the flux of gas molecules.

of the cube is unity). Since the gas is continuous, the wall is hit by a box having a volume of $\overline{v}$ per unit time (Fig. 2.6).

Therefore, when a gas has a molecular density of n and average speed of $\overline{v}$, the number of molecules that hit the unit area of the wall per unit time, or the **flux** of incident molecules, J, will be $n\overline{v}$ (here molecules $m^{-2}s^{-1}$).

In this model, one-dimensional motion of the molecules is assumed. In reality, molecules travel three-dimensionally and collide with the wall from various directions. Considering solid angle projection, the actual flux is known to be

$$J = \frac{n\overline{v}}{4} \tag{2.23}$$

By substituting Eq. (2.4) and Eq. (2.19), we obtain[10]

$$J = \sqrt{\frac{1}{2\pi mkT}}\, p \tag{2.24}$$

The molecular incident flux is very important in vacuum technologies. It is used to evaluate the number of molecules that pass

---

[10]Using constants and standard conditions, we obtain concise equation

$$J = 2.6 \times 10^{24}\, p/\sqrt{MT} \text{ (molecules m}^{-2}\text{ s}^{-1}),$$

where the unit of p is Pa, M is the molecular weight (40 for Ar).

through an opening of a wall (of a vacuum chamber, for instance) or an orifice. It is also used to evaluate the time needed by impure gas molecules to completely cover a clean surface.

2.5.2.2 Number of molecules passing through an orifice (in the case of molecular flow)

Incident flux is defined as the number of molecules passing a unit area and the molecules do not have to necessarily "collide" with a real wall. Eq. (2.23) or Eq. (2.24) is used to define an arbitral plane in space. Obviously, these equations are used for an opening in a wall (Fig. 2.7).

Let us consider the case where there is a pressure difference between the front and back sides of an opening. As gas molecules in a vacuum do not interfere with each other, Eq. (2.24) holds true for both the number of molecules moving from the high-pressure side and those from the low-pressure side. The net flux is expressed as their difference,

$$J_{1\to2} = \sqrt{\frac{1}{2\pi mkT}}(p_1 - p_2) \qquad (2.25)$$

where p_1 is the pressure of the high-pressure side and p_2 is that of the low-pressure side ($p_1 > p_2$), and therefore $J > 0$. The unit is $m^{-2}s^{-1}$. The molecular flow is the transfer of molecules from the high-pressure side to the low-pressure side, but we understand that there is always a counter from the low-pressure side. The sign of J is quite obvious, and when $J < 0$, the direction of flow is opposite and $p_1 < p_2$.

In the above equations, the "number" of molecules or atoms is used, whereas volumetric quantities are more intuitive, as we usually use volumetric units to measure the amount of gases. The amount of gas flow expressed with, for instance, ℓ per unit time is more natural and understandable to us. The amount of a gas (number of molecules or molar amount) contained in a fixed volume is dependent on the gas pressure. From Eq. (2.1), we know that the number of molecules or molar amount is proportional to the product of the pressure and the volume, at a fixed temperature.

For this reason, in vacuum technology, the product of pressure and volume is often used as the rate of gas flow. The rate of gas flow

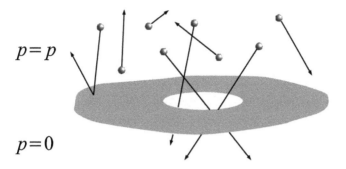

$$p = p$$

$$p = 0$$

Figure 2.7 Molecules passing through an opening.

from p_1 side to p_2 side is

$$Q = \sqrt{\frac{kT}{2\pi m}} S(p_1 - p_2) \tag{2.26}$$

This equation is obtained by multiplying kT to Eq. (2.25), which is obvious from (molecules $\times kT$) = (pressure $\times$ volume).

A typical measurement unit is Pa·m^3/s and 1 Pa·m^3/s means that a gas having a pressure of 1 Pa passing through a volume of 1 m^3 in 1 second. Some of the old gas flow meters use Torr·ℓ/s, and their readings have to be converted to SI units.

2.5.2.3 Adsorption of molecules onto clean surface

After you wash or clean or polish or scrub a dirty surface, the surface looks clean, and you are satisfied with your fine job; but physically or chemically speaking, you can never obtain a "clean surface" in such a way. From an academic point of view, a surface is considered "clean" if it atomically clean, that is, there are no atoms on the surface. Such a surface can be obtained by cleaving a (crystal) specimen in ultra-high vacuum. A clean surface is highly reactive, and the incident atoms, other than inert gas atoms, are easily adsorbed and cannot be released. Therefore, the number of atoms adsorbed in a unit area in unit time are identical to the incident flux J.

Assume that the temperature of a room is $T = 300$ K and mass of air is $M = 29$, then the incident flux is $J = 2.8 \times 10^{22} p$ (m^{-2}s^{-1}).

The areal density of atoms is roughly estimated to be 1.5×10^{19} m^{-2},[11] and therefore the time in which the incident atoms completely cover the surface is approximately $1.5 \times 10^{19}/J \approx 6 \times 10^{-4}/p$. Therefore, a clean surface is covered with incident molecules/atoms within few seconds even at 1×10^{-4} Pa (beginning of a high vacuum). Surface cleanliness can be maintained for a longer time if the vacuum is good. At 10^{-9} Pa, which is equal to the pressure 1000 km above sea level, almost one day is needed to cover the full surface!

2.5.2.4 Evaporation

Evaporation in an ideal liquid occurs at the surface below the boiling point, when the liquid changes to gas. When this process occurs at a solid surface, it is called sublimation, and the phenomenon of sublimation is included in evaporation in vacuum technologies. Condensation is the counter process that occurs at a surface when gas molecules get captured. The pressure at which the rates of evaporation and condensation are balanced (equilibrium) is called **equilibrium vapor pressure**.

When the equilibrium vapor pressure, p_e, is much lower than the vacuum pressure, the evaporated atoms fly far away from the surface and do not come back, that is, intensive evaporation takes place. The rate of evaporation is expressed by the equation of flux J. When using the flux Eq. (2.25), we set p_1 to p_e and p_2 to zero.

Vacuum drying, vacuum evaporation, and molecular beam epitaxy (MBE) are thin-film processes that use the evaporation phenomenon. We will study the theory and practical examples of evaporation-based thin film fabrication in detail in Chapter 4.

2.5.3 Cosine Law

So far, we have studied the flux of gas molecules travel freely in a space. When the gas molecules collide with a surface or when the gas molecules are released from a wall or an evaporation source, the surface confines the motion of the molecules, and therefore, the flux intensity depends on the direction of the movement.

---

[11] Here we roughly assume that the density of a solid is 10 g/cm^3 and atomic weight is 100. Please calculate.

Imagine two neighboring rooms. One room is lighted and bright, and the other room is dark. You are in the dark room. You open a small hole between the wall by offsetting the light source so that you can observe the brightness inside the lighted room. From experience, you know that the intensity of light appears strongest when you see the hole from the front, and the intensity become weaker when you see the hole obliquely. As the size of your pupil is of constant size, the total amount of light that enters your pupil decreases as the oblique angle from the wall normal increases. That is, the flux of light incident on your eyes—light intensity ÷ pupil area—is dependent only on the positional relationship between the hole and you, whereas the light flux—light intensity ÷ hole area—is not dependent on your position, as the light from the light source is scattered and the light streaming from the hole is purely diffusive (diffused light).

From this gedankenexperiment, we obtain a general rule called **cosine law**.

> Intensity of radiation from (or incident on) an infinitesimally small region, regarded as being a point, is proportional to the cosine of the angle, θ, from the normal, or $\cos\theta$.

When you see the hole obliquely, the apparent hole size actually decreases. If the area of the hole is dS when you see it from the normal, then the apparent area, that is, when you see the hole at an oblique angle of θ from the normal, decreases to $\cos\theta dS$ (see Fig. 2.8). The light intensity is proportional to the area and the decreased intensity is proportional to $\cos\theta dS$. This means that the difference between the light fluxes observed from the normal and at an oblique angle θ is proportional to $\cos\theta dS$.

As the light radiating from the point source is diffused, it expands spherically, and thus the radiation intensity is inversely proportional to the square of the distance r. The radiation intensity per unit area, or radiation flux, observed from a distance r is proportional to $\cos\theta/4\pi r^2$. The net intensity at a point source having an area of dA is

$$\frac{\cos\theta \, dS \, dA}{4\pi r^2} \qquad (2.27)$$

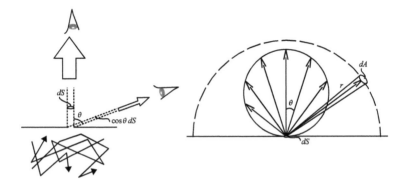

Figure 2.8 Cosine law.

Coffee Break: How Fast Do Gas Molecules Fly?

You know that the speed of sound is about 330 m/s at 0°C. The mean (root mean square) speed of air molecules, $\sqrt{\overline{v^2}}$, is about 480 m/s, which is slightly higher than the speed of sound. This is not a coincidence.

When sound waves propagate in air, it behave as a continuous medium. The reason is that since gas molecules are not bound to each other, air behaves like a bulk in which molecules collide with each other. Therefore, sound waves can never travel faster than the (mean) speed of the molecules.

The relationship between the molecule speed and the sonic speed is obtained by combining the propagation theory of sound and thermodynamics; and the speed of sound is about 70% of the speed of gas molecules ($\sqrt{(\text{heat capacity ratio of gas})/3}$).

It is important to note that dS and dA are mathematically infinitesimally small areas. This is because the above relationships are to be extended to cases where radiation sources and observation planes have a complicated geometry. In actual cases, the radiation sources and observation planes have either a two-dimensional or three-dimensional distribution shape, and the radiation flux at an arbitrary (infinitesimally small) point on the observation plane is to be obtained by adding the fluxes from an each (infinitesimally small) point on the source, that is, by carrying out multiple integration. The applications of cosine law will be presented with the practical examples.

2.6 Mean Free Path and Collision Probability

2.6.1 Mean Free Path

Gas molecules collide with each other with tremendous frequency except at very low pressures. The distance that a molecule covers between successive collisions is called the free path.

Molecular collisions are purely stochastical and the molecular speeds have a distribution; and therefore, the free path has a distribution. The average of this distribution is called the **mean free path**, frequently abbreviated as MFP. The mean free path is a crucial quantity needed to describe the behavior of gas molecules at reduced pressure. The mean free path is also closely related to viscosity and thermal conductivity of a gas.

Imagine children running around perfectly freely in a gym, and they are all blindfolded. As each of them does not know how other children are moving, they will collide with each other very frequently. Some children can run long without colliding and some children will collide very soon. If the number of the children is small (low density), they can run longer without colliding, or we can say will have a longer mean free path. Obviously, when the gym is full of children (high density), they cannot move as freely and will have a shorter mean free path. Another important measure to be considered here is the average area that one child occupies. This is obtained simply by dividing the total area of the gym by the number of children. Although they are running around, the average area of one child is constant. (In the case of gas molecules, the average volume occupied by one molecule is the inverse of the molecular density, $1/n$, as n is a number of molecules in a unit volume.)

Now we will discuss the collision area that is defined as the collision cross section. Again, we take the example of the running children. To simplify, one child is just standing, and another child runs against the standing child. So, how large will the standing child appear to the running child. They may collide nose-to-nose but that is not the only way they can collide. In another collision, they may touch on the shoulder and the running child may find the standing child twice larger!

The same is the case for the gas molecules. The collision events that take place between the gas molecules are purely stochastic when another molecule, or the target molecule, is present on the way of their travel. Therefore, for the time being, the position and motion of the target molecule does not have to be necessarily considered, and we can set that the target is at rest.

Let us now assign d as the diameter of a molecule. When the flying molecule collides with the target molecule in any way, it looks a circle whose radius is d (not diameter). This area is called collision cross section, and the area is πd^2. That is, the target looks four times larger, and the flying molecule is identified as a point. This is the same when a molecule having a diameter of d collides with a stationary point object (Fig. 2.9).

Now recall the volume that one gas molecule occupies. It was $1/n$. When a molecule travels the distance of the mean free path, λ, and collides with another molecule, the volume that it sweeps per unit

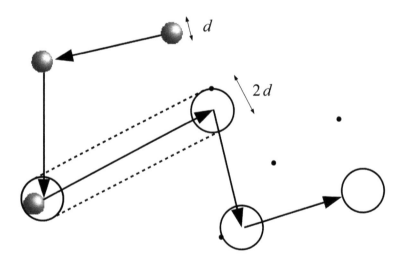

Figure 2.9 Collision events of freely moving molecules.

time is simply $\pi d^2 \lambda$ (formula for the volume of a cylinder) and this is identical to $1/n$ because all molecules move in the same manner on an average. That is,

$$n\pi d^2 \lambda = 1 \tag{2.28}$$

The number of molecules in a gas at 10 Pa is 2.65×10^{15} per cm^3, and one molecule occupies a space of 3.8×10^{-16} cm^3. Therefore, the mean free path of this gas is about 1 mm.

As the molecular number density n is proportional to the pressure, p, and is inversely proportional to the temperature, T [see Eq. (2.4)]

$$\lambda \propto \frac{T}{p} \tag{2.29}$$

The actual molecules are never at rest and moving with each other. When the target molecule is moving, we use the average relative speed $\sqrt{2}\bar{v}$ instead of the average speed $\bar{v}$ (derivation omitted). By using $\sqrt{2}\bar{v}$, we obtain a modified but more accurate equation of the mean free path

$$\lambda = \frac{\sqrt{2}}{2\pi d^2 n} \tag{2.30}$$

Practically, the mean free path estimated roughly [Eq. (2.28)] is useful enough to be employed.

By substituting values in the above equation, using d (Å), T (K), and p (Pa), we obtain λ (m) $= 3.11 \times 10^{-4} T/pd^2$. Table 2.4 lists the mean free paths at 1 Pa and 25°C (298 K) (in mm).[12] You can memorize simply that **at 1 Pa and room temperature, MFP = 1 cm**, and then the values at other pressures and temperatures can be calculated easily from Eq. (2.29).

So far we have studied that the number density, incident flux (to a wall), mean free path, and coverage time are functions of pressure. Figure 2.10 is a reckoner graph chart that summarizes these values calculated for air ($M = 29$).

---

[12] It may sound strange that an air molecule has a unique diameter, but this value is used commonly in vacuum technologies.

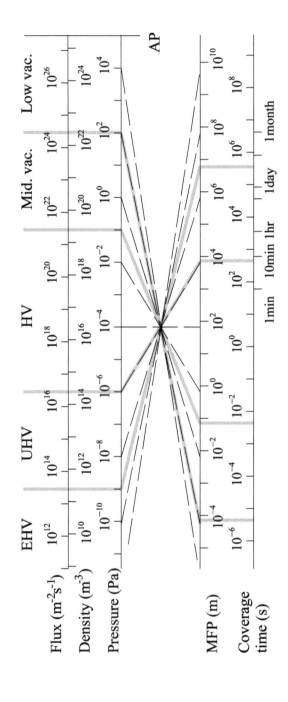

Figure 2.10 Quantities important in vacuum technology, plotted against pressure.

Table 2.4 Average molecular diameter d (Å) and mean free path at 298K and 1 Pa, λ (mm)

	H_2	He	Air	O_2	Ar	CO_2
d (Å)	2.75	2.18	3.74	3.64	3.67	4.65
λ (mm)	12.3	19.6	6.8	7.2	7.1	4.5

2.6.2 Collision Probability

Like a space ship that navigates through star dusts, a gas molecule flies through the cloud of other molecules. Some molecules may fly long distances, however, unlike the spaceship, they cannot adjust their course of travel and will collide with another molecule eventually. The longer the molecule flies, the probability to encounter an obstacle on its pathway increases—it's a mortal flight!

Let us set up a possibility that a particle survives and moves a distance x, without colliding with another particle. Again, we assume a one-dimensional model and that the target molecules are at rest. The collision cross section is $\pi d^2 n dx$ and the target molecules are distributed here and there. The longer a molecule travels, the more targets it averts (accumulation of passed target area). This means that it is losing pathways, or free space, it can travel, and will eventually collide with another particle.

If the particle travels a distance dx from x, it is now at a distance $x + dx$. The number of particles in this space is $n dx$ per unit area normal to the direction of the motion of the particle. The collision cross section of these particles is $\pi d^2 n dx$. Equation (2.28) indicates that a collision must happen (on an average) when the particle has travelled a distance λ, or when the cross section is $\pi d^2 n \lambda$. Therefore, for a travel distance of dx, the collision probability becomes dx/λ times smaller.

Now let us define the "survival" probability at position x as $P(x)$. The probability is expressed as a product of the probabilities of each event, the "collision" probability within a distance dx is $P(x)dx/\lambda$, and the survival probability decreases exactly by the same value. That is, the change in $P(x)$ is

$$dP = P(x + dx) - P(x) = -P(x)\frac{dx}{\lambda} \tag{2.31}$$

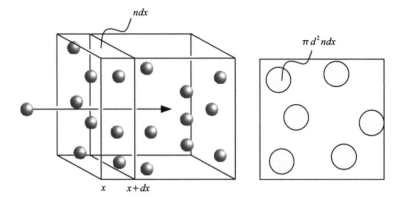

Figure 2.11 When a particle travels a distance dx, the total cross-section area is $\pi d^2 ndx$ as shown in the right.

By solving this simple differential equation under the initial condition of $P(0) = 1$, we obtain

$$P(x) = \exp\left(-\frac{x}{\lambda}\right) \tag{2.32}$$

Therefore, the survival probability decreases exponentially with the travel distance.

The collision probability should be

$$1 - P(x) = 1 - \exp\left(-\frac{x}{\lambda}\right) \tag{2.33}$$

From this analysis we know that only about 40% of the gas molecules can travel longer than their mean free path λ ($P(\lambda) = 1/e$). When a molecule travels a distance $\bar{v}$ (in one second), its collision frquency, θ, with other molecules on an average will be

$$\theta = \frac{\bar{v}}{\lambda} \tag{2.34}$$

2.6.3 Mean Free Path of a Gas Mixture

So far, we have assumed the gases to be pure, and before closing this section, we will discuss the mean free path of a gas mixture. When a dilute gas B is mixed with gas A, B molecules collide with only A molecules and A molecules also collide with only A molecules. Mean free path for A molecules are the same as we have seen above. For B molecules, the collision cross section is altered as well as molecular speeds:

$$\lambda_B = \frac{4}{\pi n_B (d_A + d_B)^2 \sqrt{1 + \dfrac{m_B}{m_A}}} \tag{2.35}$$

This can be extended to collisions between molecules as well as electrons. From $n_A \gg n_e$, $d_e \ll d_A$, $m_e \ll m_A$, we obtain

$$\lambda_e = \frac{1}{\pi \dfrac{d_A^2}{4} n_A} = 4\sqrt{2}\,\lambda_A \tag{2.36}$$

meaning that the mean free path of electrons is about 5.6 times larger than that of the molecules. Here we have discussed about electron collision because it is crucial for ionization (Section 3.3.2).

2.7 Flow of Molecules Under Vacuum

2.7.1 Viscous Flow and Molecular Flow

In this section, the dependence of gas flow on pressure is discussed. To simplify, we will assume the flow in a pipe.

It is natural that a gas will flow in a pipe if the pressure at its one end is different from its other end. The gas flows from the higher pressure end to the lower pressure end.

When the gas pressure is high then most of the gas molecules will collide with each other, and only a small member of them will collide with the wall of the pipe; that is, the diameter of the pipe, D, becomes much larger than the mean free path, λ, of the molecules, that is, $D \gg \lambda$. When this condition holds, the gas can be treated as a continuous medium and is called **viscous fluid** or having a viscous flow. The viscous fluid is described with the classical fluid dramatics. In literal sense, the pressure difference is the driving force for the gas entering a pipe. Since the gas behaves as a continuous medium, this applied force acts on the gas as soon as it enters the pipe.

When the pressure is low and satisfies $D \ll \lambda$, the macroscopic motion of the entire gas is governed by the collision between the gas molecules and the wall, rather than between the gas molecules. This type of flow is called **molecular flow**. Like the viscous flow, the gas flows from the high-pressure end to the low-pressure end, whereas the underlying mechanism is completely different. Here the mechanical force does not act and the gas molecules move completely independently without any interaction or collisions. The higher-pressure side contains more molecules (higher n) and more molecules diffuse out according to Eq. (2.23) (higher flux) as we have seen in Section 2.5.2.2. The lower pressure side contains less molecules (low n) and smaller number of molecules diffuse out (lower flux). As a result, the flux on the higher side is higher than the flux on the lower side; the net flow is from the higher side to the lower side.

In the above discussion, we set D as the diameter of the pipe. Generally speaking, it is the minimum dimension that can be considered in the system in which the gas molecules are in motion. For instance, it corresponds to the diameter of an opening, or orifice, or the height of a vacuum chamber. We call this measure **characteristic length**.

$$\text{viscous fluid}: \quad D \gg \lambda \qquad (2.37)$$
$$\text{molecular fluid}: \quad D \ll \lambda \qquad (2.38)$$

In this book, we will cover the behavior of gases under molecular regime, otherwise it will be specifically mentioned. Molecular flow is assumed in most cases of vacuum technologies. In fact, we have already assumed molecular flow when we introduced Eq. (2.25), where the diameter of the orifice is smaller enough than the mean free path of the gas.

2.7.2 Conductance

The amount of gas that passes through openings on both sides of a pipe in which a pressure difference exists, or the gas flow rate, Q, is expressed by Eq. (2.26). The proportionality coefficient is called **q**, or vacuum conductance, and is commonly denoted by C.

$$Q = C \Delta p \tag{2.39}$$

Therefore, for an opening that has an area of S and a negligible thickness,

$$C = \sqrt{\frac{kT}{2\pi m}} S \tag{2.40}$$

Vacuum chambers and vacuum pumps are connected by pipes usually made of stainless steel. The diameter of the pipes are sufficiently smaller than the mean free path of the gas molecules that travel downstream, colliding with the pipe wall repeatedly (**Knudsen** flow). In contrast to our natural belief, the collision between a gas molecule and the wall is not elastic—not like a ball bouncing off the wall—and they get **adsorbed** in the wall and lose their kinetic energy. They reside there till they gain enough thermal energy to get released (desorption). Therefore, the gas molecules "forget" their history of motion, and their desorption becomes isotropic.

Since the desorption becomes isotropic, as we have seen in the emission of gas molecules from an orifice (see Section 2.5.2.2, the amount and direction of the desorbed gases can be described by the cosine law. Therefore, the amount of a gas that moves through a pipe is obtained by executing an areal integration in three-dimensional space using equation of the cosine law [Eq. (2.27)]. The calculation is usually very complicated and the analysis is limited to simple cases. For a pipe with a radius of r and a length of L,

$$C = \frac{4}{3} \sqrt{\frac{2\pi kT}{m}} \frac{r^3}{L} \tag{2.41}$$

2.7.3 Conductance Calculus

Let us evacuate a chamber with a vacuum pump through a pipe connected to the chamber. By setting the chamber pressure to p and the pressure of the entry of the pump to p_p, the gas flow rate through the pipe is

$$Q = C(p - p_\mathrm{p})$$

where C is the conductance of the pipe. If another pipe is connected in parallel, the flow rate is doubled obviously (assuming that the

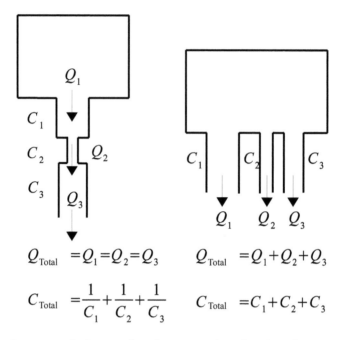

Figure 2.12 Definition of conductance and combined conductance.

pump functions ideally). From this, we know that the parallel sum of conductance is simply

$$C_{\text{Total}} = C_1 + C_2 + C_3 + \cdots \tag{2.42}$$

The series sum of conductance, when vacuum components are connected in series, can be easily found by

$$\frac{1}{C_{\text{Total}}} = \frac{1}{C_1} + \frac{1}{C_2} + \frac{1}{C_3} + \cdots \tag{2.43}$$

because Q through the vacuum components is constant.

The equation of the series sum corresponds to the parallel sum of electric resistance, and the equation of the parallel sum corresponds to the series sum of electric resistance. Conductance is defined as the inverse of resistance, and therefore, the above equations are actually identical to those in electric circuits.

2.7.4 Vacuum Flow Rate and Pumping Speed

Flow rate of evacuating gas Q is called **vacuum flow rate** and it is usually measured with a unit of $Pa \cdot m^3/s$.

The performance of vacuum pumps is measured by the **pumping speed** S, which is a quantity obtained by dividing the vacuum flow rate (throughput) Q by the then pressure p_p (usually at the entry of the pump)

$$S = \frac{Q}{p_p} \qquad (2.44)$$

Common units of the pumping speed are m^3/s and ℓ/s. Obviously the pumping speed indicates the gas volume exhaust per unit time.

Assume that a vacuum pump has a performance of 1000 ℓ/s regardless of pressure. The volume of gas exhausted per unit time is always constant as understood from the unit of the pumping speed. However, the amount (mass or number of molecules) of gas decreases as the pressure decreases.[13]

The pumping speeds of rotary pumps, oil diffusion pumps, and turbo molecular pumps are almost constant under a wide range of operating pressures. This is obvious for rotary pumps and turbo molecular pumps because the volume "scooped out" by one rotation of blades is constant.

Now we will discuss a vacuum chamber that has been evacuated at a constant pumping speed S through a pipe with conductance C. The vacuum flow rate Q is

$$Q = Sp_p = C(p - p_p) \qquad (2.45)$$

Now we focus on the pressure p at the pipe entrance—that is identical to the pressure of the chamber. **Effective pumping speed** S_{eff} is

$$S_{eff} = \frac{Q}{C} = \left(\frac{1}{S} + \frac{1}{C}\right)^{-1} \qquad (2.46)$$

---

[13]When we measure the gas flow rate, a dimension of volume divided by time is usually used. This unit is definitely same as that of the pumping speed, whereas the amount of gas flowing per unit time may not be the same. While measuring the "gas flow rate" the temperature and pressure are usually assumed to be constant, and therefore, the volumetric flow rate and the mass flow rate are identical, which is not very obvious to beginners. To avoid confusion, refer to Section 2.5.2.2.

and this quantity is always lower than the pumping speed S of the pump.

Therefore, when a vacuum chamber and a vacuum pump are connected with vacuum components such as a pipe, the effective pumping speed never reaches the pumping speed of the pump. When a low C component (such as a long, narrow pipe) is used, the effective pumping speed is determined by C, even if a high-performance (high S) pump is connected.

Now we evacuate a vacuum chamber that has a volume of V (m^3) at a pressure of p (Pa) with an effective pumping speed of S_{eff} (m^3/s). This means that we "scoop" a gas having a pressure p by a volume of S_{eff} (m^3) every second. Therefore, for an infinitesimal time dt, the number of molecules expelled from the chamber is

$$\frac{p S_{\mathrm{eff}}}{kT} dt$$

As a result, the chamber pressure is reduced by dp, and the number of molecules in the chamber is decreased by $V dp/kT$. As these two quantities are identical, we obtain the following differential equation:

$$\frac{dp}{dt} = -\frac{S_{\mathrm{eff}} p}{V} \tag{2.47}$$

By setting the initial pressure to p_0 and the achieving pressure to p_{eq}, we understand that the chamber pressure p decreases in an exponential manner with time.[14]

$$p = p_0 \exp\left(-\frac{S_{\mathrm{eff}}}{V} t\right) + p_{eq} \tag{2.48}$$

Actual pressure change during vacuum pumping does not follow the above equation due to the pressure dependence of pumping speed, vacuum leakage, and gas emission from vacuum components and chambers, but shows good agreement in a high-pressure (low vacuum) range.

The chamber achieving pressure is

$$p = \frac{Q_{\mathrm{L}}}{S_{\mathrm{eff}}} + p^* \tag{2.49}$$

---

[14] Therefore, for a log-scaled analog pressure meter, which is currently not very common, the needle moves at a constant speed during evacuation.

where Q_L is the gas leakage range and p^* is the ultimate pressure of the pump.

2.7.5 Gas Admission, Pressure Regulation, and Average Residence Time

In many dry processes, the pressure of a process chamber is to maintained at a constant value by admitting a process gas.

Let us examine a vacuum (low-pressure) process where the chamber having a volume of 50 ℓ is maintained at 1 Pa by admitting a process gas at a flow rate of $Q = 100$ sccm.[15]

A flow rate of 100 sccm corresponds to about 170 ℓ/s at 1 Pa. To maintain a constant pressure, the flow rate of the incoming gas and the exhaust rate (effective pumping speed S_{eff}) should be balanced, therefore, the required capacity of the vacuum system is $S_{eff} = 170$ ℓ/s.

As understood from Eq. (2.46), a pipe between the chamber and the pump makes the effective pumping speed S_{eff} several times smaller than the capacity of the pump. Therefore, we use a vacuum pump having a capacity, let us say, ten times larger than the above S_{eff}. As a result, the effective pumping speed becomes larger than the required value ($S_{eff} > 170$ ℓ). We add a vacuum adjusting valve such as a butterfly valve to reduce and adjust the conductance.

A part of the admitted gas will go through the chamber very fast, and another part of the gas will reside in the chamber longer. The average residence time, τ, is the mean time calculated when all of the admitted gas is residing in the chamber. This is simply expressed as the ratio of the chamber volume V divided by the volumetric flow rate of the gas at the working pressure (= pumping speed).

$$\tau = \frac{V}{S} = \frac{pV}{Q} \qquad (2.50)$$

---

[15] sccm: standard cubic centimeter per minute, or, the gas volume (ml) flowing in one minute converted to the standard state; practical engineering unit often used in dry processing.

The average residence time τ for the above case is 0.3 s. The average residence time is necessarily important to discuss the chemistries proceeding in the chamber.

2.8 Vacuum Equipments

The final section of this chapter is devoted to describing how to make a vacuum for micro- and nanofabrication processes. A minimum set of a vacuum system consists of vacuum chambers, vacuum pumps, pipes, and vacuum gauges. Auxiliary components such as a gas metering system, heaters, and plasma generators are added depending on the purpose of the process.

2.8.1 Vacuum Chamber and Pipe/Fitting

A vacuum chamber is a container designed strong enough to bear the atmospheric pressure. Many of vacuum chambers are made of stainless steel because it has high strength, does not rust or alter, does not emit impure gasses, and is rather well machinable and weldable. Usual steels are not very common, as the surface rust, if formed, can easily adsorb (and desorb) impure gases. Recently, aluminum materials have been developed for UHV applications and are being widely used.

Bell jars have been widely used as a vacuum chamber for a very long time. A bell jar is bell-shaped, usually made of glass, but some are made of metal. Round shoulders of the bell shape eases the stress concentration and thus allow the use of very large glass chambers.

Metal vacuum chambers are equipped with flanges and ports to which pipes, fittings, glass-view ports, and sensors are connected. Doors and lids for maintenance or sample transfer are also attached to a flange.

Vacuum pipes and fittings are also made of stainless steel or aluminum. Vacuum flexible hoses and bellows are available to form a desired shape. Hoses/bellows made of stainless steel are commonly used for high-vacuum applications. Rubber hoses are still widely used; the hose is designed thick enough to resist the atmosphere pressure.

———— Coffee Break: Spacecraft Near-at-Hand ————

Vacuum chambers and spacecraft are identical but just opposite. Inside the vacuum chamber, there is vacuum and outside it is atmosphere, whereas the inside of the spacecraft is atmosphere and the outside is vacuum. Both the vacuum chambers and spacecraft are made resistant to the atmospheric pressure, gas impermeable, and have windows and gateways. Small, space-saving, and high-performance "vacuum" components were needed and developed in space industry. Many of advanced vacuum components widely used at present in the semiconductor industry have their origins to the space components.

2.8.2 Vacuum Pumps

2.8.2.1 General classification

Vacuum pumps are generally categorized into gas transfer pumps and entrapment pumps. The gas transfer pump compresses the gas at the inlet port and then exhausts it to the outlet port. This principle is the same as the bicycle pump that compresses the air and delivers it to a tube.

The gas transfer pumps are further categorized into **positive-displacement pumps** and **kinetic vacuum pumps**. Representative positive-displacement pumps are **rotary mechanical pumps** and **roots pumps**. Representative kinetic vacuum pumps are diffusion pumps and turbo molecular pumps.

The **entrapment pumps**, literally, entrap and keep the gas inside them. Cryo pumps and sputter-ion pumps are popular.

The main specifications of vacuum pumps are pumping speed and ultimate pressure. The operating pressure range is the difference between the highest pressure (lowest vacuum) at which the pump starts to operate and the ultimate (lowest) pressure, and the

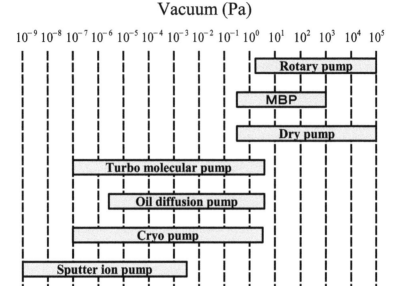

Figure 2.13 Operating pressure ranges of various types of vacuum pumps.

wider is this difference, the better. Figure 2.13 shows operating pressure ranges for different vacuum pumps.

2.8.2.2 Types of vacuum pumps

Rotary pump A rotary pump or rotary vane pump is probably the most common vacuum pump. It functions from the atmospheric pressure down to about 10^0 Pa in the case of single-stage usage. Figure 2.14 is an example of a rotary vane pump. Two vanes attached to a rotor at the center are pressed to the inner wall of a cylinder. As the axis of the rotor is offset from the cylinder axis, the volume of the space bounded by the vanes, the rotor, and the cylinder changes as the rotor rotates. Gas enters when this space has a larger volume, gets compressed with rotation, and is finally exhausted to the outlet port thorough a gas ballast valve. The internal space, such as the cylinder, is filled with oil. The oil serves to seal the outlet valve from leaks and also lubricates the moving parts. The capacity of rotary pumps for laboratory experiments (as well as for semiconductor

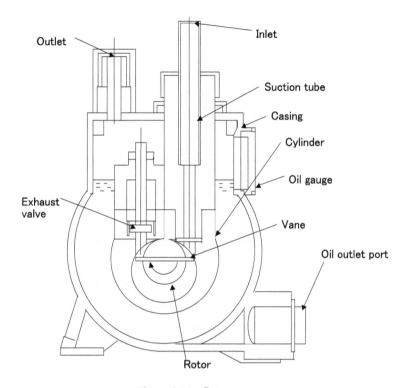

Figure 2.14 Rotary pump.

manufacturing) have a pumping speed of 50–2000 ℓ/min. Rotary pumps are economically affordable, irrefrangible, easy-to-use or -to-maintain, and function from the atmospheric pressure so that no roughening/backing pumps are necessary. For these reasons, rotary pumps are widely used. One serious drawback is oil emission. As the gas directly gets in contact with oil, the oil vapor is easily incorporated to the vacuum which limits the achieving pressure and leads to contamination. The oil must be exchanged regularly as the oil degrades, but usually not so often.

Roots pump Roots vacuum pump, also known as mechanical booster pump has one set of two cocoon-shaped rotors. Each rotor rotates to the opposite direction synchronously, and compresses and transfers the gas. The roots pumps performance best in the middle vacuum range where the pumping speed of rotary pumps drops.

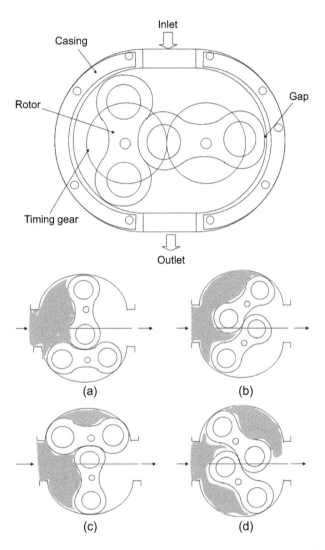

Figure 2.15 Structure and operation schemes of roots pump (mechanical booster pump).

The clearance between the rotors is very small (0.1–0.3 mm) so that the lubricant is not necessary, thus there are lesser chances of oil contamination. The capacity of roots pumps for laboratory experiments have a pumping speed of 1000–10,000 ℓ/min.

Dry pump Dry pump is the generic name of pumps that compress and transfer the suctioned gas without using lubricant or sealing oils. Most of the latest common dry pumps are used as a replacement of rotary pumps, functioning from the atmospheric pressure with a high throughput. Generally speaking, the achieving pressure is better than that of rotary pumps, providing a wider operating pressure range. Various types of dry pumps for this usage are available in market, such as roots type (described above), double rotating screw type, turbo fan type that is similar to the turbo molecular pumps shown afterward, and scroll type in which the suctioned gas is compressed and transferred with two scrolled walls (stator and rotor).

Since there is no use of oil, contamination in vacuum chambers due to the oil is avoided, and this is extremely important in micro-/nanoprocessing where ultra-cleanliness is mandatory. However, the mechanical complexity makes the price and maintenance cost much higher than those of rotary pumps.

Diffusion pump A diffusion pump (Fig. 2.16) is a type of kinetic vacuum pump. It has a very simple construction with no motion mechanism and is used for creating a high vacuum.

Diffusion pumps have a cylindrical vessel with a small volume of oil placed at the bottom. The bottom part is heated to vaporize the oil. The plum of oil rises through chimney pipes placed inside the pump cylinder. The oil vapor particles collide with the backside of the umbrellas at the top of chimneys, and then blast out forming a downward supersonic jet. Gas molecules incoming "incidentally" to the inlet port fly randomly; but some of the gas molecules are hit by heavy oil molecules and gain downward momentum. As a result, the number of molecules that move downward becomes larger than those moving upward, and a net downward gas flux is generated, meaning that the incoming gas is exhausted to the outlet port.

From this principle, we understand that the diffusion pump functions in the molecular flow range. As the upper part of the vessel is water-cooled, the oil vapor liquefies and then returns to the bottom oil reservoir. Usually 3 or 4 umbrella nozzles are equipped. The usual upper limit of the operating pressure is approximately 10^1 Pa, and achievable pressure is 10^{-6} Pa. The capacity of diffusion

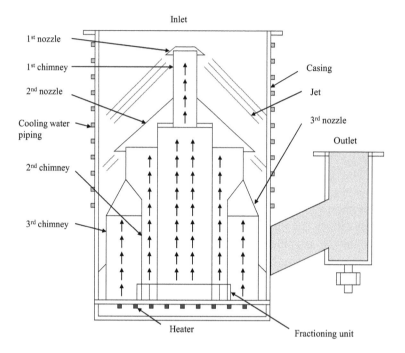

Figure 2.16 Diffusion pump.

pumps for laboratory experiments have an inlet diameter of 10–30 cm and their pumping speeds are 300–3000 ℓ/s.

Diffusion pumps can make quite high vacuum, have a very simple construction, are therefore very affordable and do not require frequent maintenance actions if operated properly. One major concern is oil contamination. Liquid nitrogen oil traps are generally used, but they are not always enough to satisfy the cleanliness requirements in industries involved in the development of recent semiconductors. Another issue is probably rather longer boot-up/shut-down time required for heating and cooling the oil reservoir.

Turbo molecular pump Fins of a turbine rotor rotating at 20,000–30,000 rounds per minute move at a velocity of several hundred meters per second. This velocity is as high as that of gas molecules diffusing between the fins, which allows transferring of the

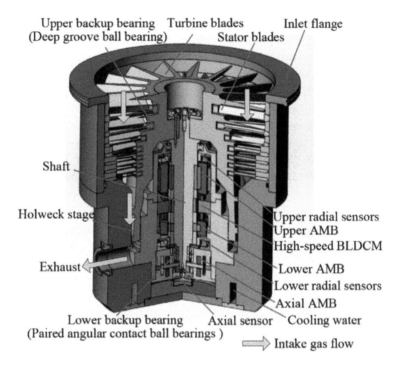

Upper backup bearing   Turbine blades   Inlet flange
(Deep groove ball bearing)   Stator blades

Shaft

Holweck stage

Upper radial sensors
Upper AMB
High-speed BLDCM

Exhaust

Lower AMB
Lower radial sensors
Axial AMB

Lower backup bearing   Axial sensor   Cooling water
(Paired angular contact ball bearings)

Intake gas flow

Figure 2.17 Turbo molecular pump. Reprinted from B. Han, et al., Design aspects of a large-scale turbo molecular pump with active magnetic bearings, *Vacuum*, **142**, p. 97, Copyright 2017, with permission from Elsevier.

downward momentum of the fins to the gas molecules, or exhausting the gas, as in the case of diffusion pumps. A stator beneath the rotor has fins aligned symmetrically to the rotor fins, the construction of which suppresses the backward flux of the gases. From this operating principle, it is obvious that the turbo molecular pumps, as well as diffusion pumps, function under a molecular flow range, but under a wider range of pressure such as from 10^{-1} Pa to below 10^{-7} Pa. Turbo molecular pumps have a complicated mechanics and are expensive, but they have as easy as one-touch operation and achieve clean, high vacuum, which makes them the standard high-vacuum pumps that have kicked out diffusion pumps from laboratories. The capacity of laboratory turbo pumps range from 50–2000 ℓ/s.

Cryo pump and sputter-ion pump   Both cryo pump and sputter-ion pump are entrapment pumps and allow to attain better vacuum than diffusion pumps and turbo molecular pumps. As cryo pump and sputter-ion pump do not exhaust gas in the usual way, these pumps cannot be used in experimental and manufacturing processes in which a process gas is admitted continuously to a chamber.

A cryo pump condenses and traps gas molecules with a cooling fin, called the cryo panel, or activated carbon cooled at an extremely low temperature (ca. 20 K). Liquid He is used as a coolant. Low-mass gases, such as H_2, He, and Ne have a lower liquefaction temperature than the panel temperature, but can be trapped physicochemically in the small pores of the activated carbon or those formed by their own molecules of the adsorbed gas. The heat radiation from the wall decreases the efficiency of the cooling panel; to prevent this, a wall cooling with liquid N_2 is recommended. The pumping speed decreases as the amount of gas entrapment increases. For this reason "regeneration" is carried out periodically to evaporate the trapped gases. Most of cryo pumps are designed to carry out the regeneration by isolating the vacuum chamber. The pumping speeds are very large; 1000–10,000 ℓ/s even for laboratory equipment. Cryo pumps are the most expensive among entrapment pumps but have become popular due to their high pumping speed and the capability of regeneration.

A sputter-ion pump traps impurity (residual) gases through chemical reactions between the ionized gases and reactive gases. The electrons emitted from a stainless steel cathode ionize the impurity gases, and the ions are accelerated by the electric field between the cathode and the anode, and collide with the anode that is made of titanium, leading to the sputtering of titanium. As raw titanium is very reactive, the sputtered titanium gases easily react with the impurity gases, form non-volatile titanium compounds, and the formed compounds deposit on the wall of the pump casing; as a result, the impurity gases are trapped. A typical pumping speed is in the range of 50–500 ℓ/s. A titanium sublimation pump, sometimes called a getter pump, is based on the same chemical principle of gas entrapment, but the titanium is thermally sublimated. Although these reactive entrapment pumps are not very economically affordable, they provide an ultra high vacuum without any motion mechanisms and therefore shows high reliability. For these reasons, the sputter titanium entrapment pumps are widely used for analysis equipment such as electron microscopes. Sputtering phenomenon will be discussed later in Section 4.3.

2.8.2.3 Introduction to practical designing of vacuum pumping systems

Figure 2.18(a) is a simple, basic, high-vacuum pumping system, using a turbo molecular pump in combination with a rotary pump. The turbo molecular pump can be replaced with a diffusion pump, and the rotary pump can be replaced with a compatible dry pump; the latter system configuration is becoming very popular. Although high-vacuum pumps, such as turbo molecular pumps and diffusion pumps, have a high pumping speed, they do not function at low inlet pressures and, therefore, the pressure at the outlet port must be maintained at a low–medium pressure (usually 10–100 Pa). To satisfy this condition an additional pump is connected in series with the high-vacuum pump. This additional pump is called a forepump or backing pump; the rotary pump in Fig. 2.18(a) is a forepump. The additional role of the forepump in the case of Fig. 2.18(a) is to reduce the chamber pressure as low as to operate the high-vacuum pump. For this purpose, a bypass line is added to connect the chamber directly to the inlet port of the forepump. This is because high-vacuum pumps do not function at high pressures, and if this done, the pumps will crash. The valve at the bypass line is opened during the initial evacuation and then closed. The low-vacuum pump used for this purpose is called a roughening pump; in this case here, the roughening pump and the forepump are identical. The system configuration—numbers and positions of valves, pipes, and pumps—may vary for each, and all the valves and pumps must be operated properly.

The higher the pumping speed, the better, but a combination of a high-vacuum pump and a forepump must be considered carefully. Using a low-performance forepump decreases the performance of the high-vacuum pump, and using a too large forepump is not cost effective.

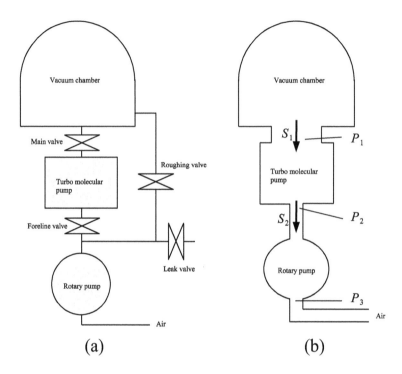

Figure 2.18 Typical vacuum pumping systems.

In Fig. 2.18, the inlet pressure of the turbo molecular pump is p_1 (assuming it to be the same as the chamber pressure), the inlet pressure of the rotary pump is p_2 (assuming it to be the same as the outlet pressure of the turbo molecular pump), the outlet pressure of the rotary pump is p_3 (same as the atmospheric pressure), and $p_1 \ll p_2 \ll p_3$ holds. The same gas flows through each pump, therefore, from the continuity law, we understand that $p_1 S_1 = p_2 S_2$ holds [Fig. 2.18(b)].

Vacuum pumps should be operated within a proper (inlet) vacuum range in which they perform the best. Within this range, the pumping speed is almost constant, and it decrease outside this range. A graph chart that depicts the relationship between the pressure and the pumping speed is a **performance curve**. Figure 2.19 shows examples of pump-performance curves.

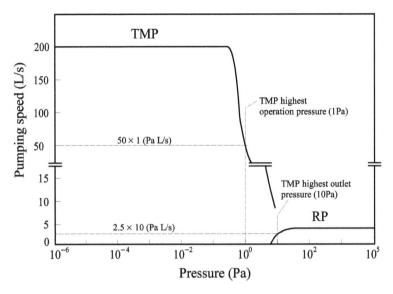

Figure 2.19 Performance curves for turbo molecular pump (TMP) and rotary pump (RP). Vertical scale is not continuous throughout.

Performance curves allow us to design a proper combination of a high-vacuum pump and a forepump. Figure 2.19 shows a combination of a turbo molecular pump with $S = 200$ ℓ/s and a rotary pump with $S = 4$ ℓ/s (240 L/min).

When evacuation is switched from roughening pumping (rotary pump only) to main pumping (rotary pump plus turbo molecular pump), the amount of the residual gas in the vacuum chamber is still high because the initial gas flow rate through the turbo molecular pump is very high and the pump is highly loaded. The chamber pressure at this occasion is usually as low as the highest operating pressure of the turbo molecular pump; it is about 1 Pa in the case of the pump shown in Fig. 2.19, and the gas flow rate (throughput) is estimated to be $50 \times 1 = 50$ Pa·ℓ/s.

The forepump should pump this gas flow. Assuming the worst case that the outlet pressure of the turbo molecular pump is as high as the critical backpressure (highest outlet pressure) allowed, this value is 10 Pa for this case, and from the relationship shown above, $p_1 S_1 = p_2 S_2$, the required specification of the forepump is to be more than $50/10 = 5\,\ell/s$ ($>120\,\ell/\min$). The rotary pump shown in Fig. 2.19 adequately satisfies this requirement.

Exercises

(1) Obtain the molecular number density at 0 Pa and 500 K.

(2) A mixed gas consists of 0.1 mol of N_2 and 0.2 mol of O_2.

  (a) Calculate its volume and molecular number density at 1.013×10^5 Pa and 500 K (1.013×10^5 Pa is identical to 1 atm).

  (b) Calculate the partial pressure of O_2, when this gas is compressed to 1 ℓ.

(3) Derive the "ideal gas law at molecular level," $p = nkT$, from the usual ideal gas law, $pV = NRT$.

(4) In thermal motion, speeds of particles, such as gas molecules, distribute from zero to infinity following a mathematical function.

  (a) What is the name of this function?

  (b) Using this function, calculate the most probable velocity that the gas molecules will have.

(5) Derive the dimension of each parameter in Eqs. (2.4), (2.9), and (2.12). Use L for length, M for mass, T for time, and K (Kelvin) for temperature.

(6) Derive the equation for the relationship between pressure and speed of gas molecules.

(7) An Ar atom has a kinetic energy of 0.1 eV. Calculate the speed of this atom.

(8) Calculate the mean free path of an Ar gas at 10 Pa and 500 K.

(9) Derive Eq. (2.18) from Eq. (2.15).

(10) A gas is flowing at 1 Pa·m^3/s. How many molecules are moving per second? Convert this value to mol/s. Temperature is 300 K.

(11) Explain the following terms:
distribution function of speeds, Maxwell–Boltzmann function, mean free path, cosine law, rotary pump, diffusion pump, turbo molecular pump, entrapment pump

(12) A vacuum chamber is connected to two rooms with through small orifices on the walls. The area of each orifice is S. The pressure of one room is p and the pressure of another room is sufficiently low compared to p. Assume a point on a plane parallel to the wall. The distance between this plane and the wall is r. Obtain the incident flux at a point at an angle of θ to the normal of the orifice. The wall thickness and the reverse flux are negligible. The orifice is negligibly small in terms of geometrical consideration.

Chapter 3

Fundamentals of Plasma

3.1 Introduction

Plasmas are used in the formation and patterning of thin films. A plasma is a gas involving electrically charged species of ions and electrons. The motion of electrically charged species can be controlled by an electric field or a magnetic field. This means that a gas in a vacuum environment can be controlled, even though only a part of it can be controlled. Although a part of the gas is controllable, but the motion of the gases cannot be controlled by classical gas technologies. Another important aspect of using plasma is to generate highly chemically reactive molecules (radicals), which assist in proceeding chemical reactions better than by usual thermal processes. Therefore, the use of plasma realizes material fabrication, which is not straightforward by conventional techniques, that are, the gas motion controlling and the chemical reaction enhancement. Therefore, we can use gas molecules as a "tool" in micro- and nanoprocesses (Fig. 3.1).

Micro- and Nanofabrication for Beginners
Eiichi Kondoh
Copyright © 2021 Jenny Stanford Publishing Pte. Ltd.
ISBN 978-981-4877-09-1 (Hardcover), 978-1-003-11993-7 (eBook)
www.jennystanford.com

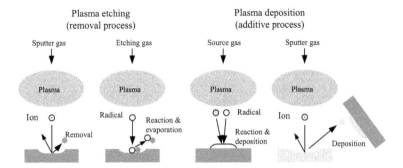

Figure 3.1 Principles of plasma microfabrication.

Roles of plasma in micro-/nanofabrication

(1) To generate ions and use them as high-energy particles.
(2) To generate radicals and use them to assist in chemical reactions.

In this chapter, we will study the definition and properties of plasma and the motion of charged particles inside a plasma. We will also study the generation and internal structures of plasma.

3.2 What Is a Plasma?

3.2.1 Particles in a Plasma

A **plasma** is defined as a gas in which freely moving ions and electrons coexist, keeping the net electrical charge neutral. Bolt of lightning, aurora, gases inside fluorescent lamps and neon signs when they turn on, the inside of stars, and the interstellar medium are all examples of a plasma. A good guesser will notice that plasma is related to the (electrical) discharge of a gas.

Gas discharge or simply "discharge" is a phenomenon in which electric current flows in a gas. This occurs when a gas contains charged species, such as ions and electrons; otherwise gases are non-conductive, because the gas molecules are electrically

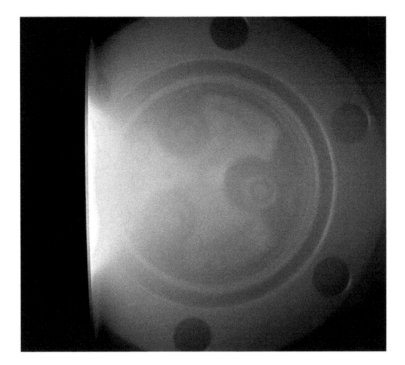

Figure 3.2 Plasma discharge (DC sputter).

neutral. Plasmas can be easily generated thorough a discharge phenomenon.

In a plasma, a part of a gas is ionized (**ionization**), and positive ions, electrons, and **neutral particles** are generated. The discharge of a gas is identical to ionization in which neutral atoms and molecules dissociate into positive ions and electrons.

These ions are usually monovalent, or singly ionized, therefore, the number of the ions and the electrons is the same. The neutral species are not charged, and we can presume that they are identical to the particles that are usually regarded as gas molecules. Elements that have a high electron negativity can exist as negative ions in plasmas, which is however usually not considered in introductory textbooks. In this book, only positive ions are assumed.

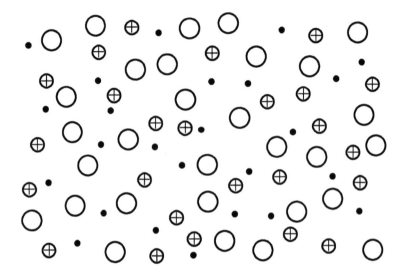

Figure 3.3 Conceptual depiction of particles in plasma.

The ratio of the number density of electrons n_e to that of neutral particles n_n

$$\frac{n_e}{n_n} \qquad (3.1)$$

is called **degree of discharge**, and n_e is called **plasma density**. A plasma in which the degree of discharge is roughly $n_e/n_n < 10^{-3}$ is called a **weakly ionized plasma**, and this book describes only this type of plasma.

Figure 3.3 depicts the particles in a plasma conceptually, where the number of the ions and the electrons is the same, and a majority of particles are neutral molecules. Table 3.1 lists representative physical values of the particles in a low-temperature plasma.

Ions and electrons move freely in a plasma. In this sense, either ions are electrons or a "gas," and their kinetic energy and temperature are described by Eq. (2.12):

$$\frac{1}{2}m\overline{v^2} = \frac{3}{2}kT \qquad (3.2)$$

Table 3.1 Characteristics values of a typical low-temperature plasma $(n_i = n_e = 10^{16} \text{ m}^{-3})$

	Ion	Electron	Neutral particle
Mass	$m_i = 6.6 \times 10^{-26}$ kg	$m_e = 9.1 \times 10^{-31}$ kg	$m_n = m_i$
Temperature (energy)	500 K (0.04 eV)	23200 K (2 eV)	293 K (1/40 eV)
Mean motion speed	$\overline{v_i} = 5.2 \times 10^2$ m/s	$\overline{v_e} = 9.5 \times 10^5$ m/s	$\overline{v_n} = 4 \times 10^2$ m/s
Incident current	0.21 A/m^2	380 A/m^2	—

The coefficient on the right hand side (3/2) is related to the space of motion, or the degree of freedom, and is determined by the motion model of concern. A more important thing is that the kinetic energy is roughly identical to kT, and kT can be frequently found in physical laws and formulas related to the phenomena of thermal motion. Therefore, we can simply define the conversion between thermal energy and temperature as

$$kT \tag{3.3}$$

It can be noted that Eq. (3.3) is directly obtained by substituting the most probable speed $v_m = \sqrt{\frac{2kT}{m}}$ into Eq. (3.2). It is not our subject to judge whether the former or the latter equation is correct, but it is more important to notice that kT is a guide of thermal energy.

Usually, in a plasma employed for plasma processing, electrons have a much higher average speed than the other species. Therefore, the electron temperature is much higher than the ion temperature. A plasma that has different ion and electron temperatures is called **non-equilibrium plasma**. **Low-temperature plasma** is also another name for this type of plasma, because the ion temperature is lower than that of the electron temperature.

Let us now look at the unit of energy that is conveniently used for the kinetic energy of electrons. It is electron volt (eV). 1 eV is the energy obtained when a unit charge is accelerated through

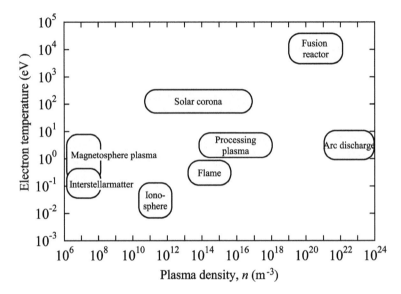

Figure 3.4 Relationship between the plasma density and the energy of various plasmas.

an electrical potential difference of 1 V and satisfies the following relationship:

$$1 \text{ eV} \equiv 1.60219 \times 10^{-19} \text{ J} \qquad (3.4)$$

1 eV roughly corresponds to about 10,000 K. As shown in Table 3.1, the electron energy in a plasma is approximately 2 eV, equivalent to about 20,000 K temperature. As room temperature (300 K) is equivalent to 25.9 meV, the electron temperature is surprisingly high. On the other hand, the ion temperature is higher than room temperature by only 100–200 K, and the temperature of neutral particles is essentially at room temperature.

Figure 3.4 shows the relationship between the plasma density and the energy of various plasmas.

Coffee Break: How High is 1 eV?

Electron volt is usually used to measure kinetic energy and interaction of small particles such as electrons, atoms, and molecules. As stated earlier, a thermal energy of 1 eV is equivalent to about 10,000 K in temperature! Be that as it may, we cannot easily imagine how high is it.

Water vaporizes at 100 °C, and its heat of vaporization is 2260 J per 1 cm^3 (40.66 kJ/mol), equivalent to 0.43 eV per molecule. The energy of combustion of propane (C_3H_8)—propane gas used for boiling water here—is 23.0 eV and 0.23 eV per chemical bond. The voltage of a dry cell is 1.5 V, because this electromotive force is arising from the electrochemical reactions inside the cell.

As seen above, familiar chemical and physical phenomena have an energy within an order of magnitude of about 1 eV. In other words, the thermal energy of room temperature is too small. All the substances around us are mostly chemical compounds and are substantially stable—our body as well. This means that all the usual chemical reactions around us proceed gently. The thermal energy at room temperature is not enough to lead to a catastrophic chemical phenomena. Therefore, our present world is quite stable.

3.2.2 Motion of Charged Particles in an Electromagnetic Field

Plasmas for micro- and nanofabrication are generated by an electric or magnetic field. Charged particles such as ions and electrons are in motion in a plasma, and these charged species are under the influence of the electromagnetic field. The motion of the particles is deeply related to the internal physics of a plasma.

3.2.2.1 Motion in electric field

A positively charged particle moves toward a negative electrode (along the positive direction of the electric field), and a negatively charged particle moves toward a positive electrode (along the

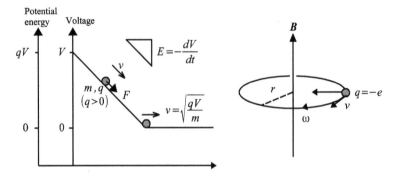

Figure 3.5 Motion of a particle in an electric field and cyclotron motion in a magnetic field.

negative direction of the electric field). This is because Coulomb force acts at a charged particle in an electric field.

The force that acts on a particle of mass m and charge q in a electric field $\mathcal{E}$ is $F = q\mathcal{E}$, and the equation of motion is

$$m\frac{dv}{dt} = q\mathcal{E} \tag{3.5}$$

Therefore, the acceleration that works on the particle is

$$\alpha = \frac{dv}{dt} = \frac{q\mathcal{E}}{m} \tag{3.6}$$

When a particle with a unit positive charge "falls" in the voltage difference of V for a very short time, Δt, by assuming a constant electric field of $\mathcal{E} = -dV/dx$, the increments of position and time are (Fig. 3.5)

$$\Delta v = \int \alpha dt = \frac{q\mathcal{E}}{m}\Delta t \tag{3.7}$$

$$\Delta x = \iint \alpha dt dt = \frac{1}{2}\frac{q\mathcal{E}}{m}(\Delta t)^2 \tag{3.8}$$

Therefore, the lighter the mass of the particle, the faster and longer the particle moves, as listed in Table 3.2. (Note that the falling time, falling distance, and the final velocity are independent of mass in the case of free fall of a particle under the sole influence of gravity. This is because gravity acts on every object equally.)

Table 3.2 Comparison of the motion of particle in an electric field and in a gravitational force field ($x = 0$ is the fiducial position)

	Gravitic	Electric
Acting force, F	mg	$q\mathcal{E} = -q\dfrac{dV}{dx}$
Acting acceleration, α	g	$\dfrac{q\mathcal{E}}{m}$
Potential energy, E, at x	$\displaystyle\int_0^x mgdx = mgx$	$\displaystyle\int_0^x q\mathcal{E}dx = -\int_V^0 qdV = qV$
Final velocity after "fall" by distance x	$\sqrt{2gx}$	$\sqrt{\dfrac{2qV}{m}}$
Final velocity after "fall" for time t	gt	$\dfrac{q\mathcal{E}t}{m}$
Kinetic energy gained after "fall" for time t	$\dfrac{m(gt)^2}{2}$	$\dfrac{(q\mathcal{E}t)^2}{2m}$

3.2.2.2 Conservation of energy

The potential energy, E, gained by a charge, q, after a "climb" by a voltage difference, V, is

$$E = qV \tag{3.9}$$

which is the work done by the electric field (Table 3.2). The word "climb" is used here with the image of gravitational potential energy in mind.

In the case of the "fall" discussed above, the particle loses its potential energy and gains kinetic energy, and from

$$qV = \frac{1}{2}mv^2 \tag{3.10}$$

we obtain the final velocity of $v = \sqrt{2qV/m}$.

The kinetic energy that a charged particle obtains after a "fall" in an electric field, $\mathcal{E}$, for a given short time Δt is equal to the electric potential energy the particle loses.

$$E = qV = qx\mathcal{E} = q\frac{1}{2}\frac{q\mathcal{E}}{m}(\Delta t)^2\mathcal{E} = \frac{(q\mathcal{E}\Delta t)^2}{2m} \tag{3.11}$$

This means that a lighter particle gains larger kinetic energy per unit time. (A heavier object gains a larger kinetic energy during a gravitational free fall.)

The mass of an electron is extremely smaller than that of an ion, and thus an electron gains much higher kinetic energy and moves much faster. Therefore, we now understand that

> Electrons in a plasma have extremely large kinetic energy (a few eV to 10 eV) and move faster than ions or neutral particles.

The mass of an ion is approximately 10^5 times larger than that of an electron, whereas the energy of an electron is only 10^2 times larger than that of ions (see Table 3.1). This is because the electron energy is transferred to slower particles upon collision with the particles. Therefore, the temperature of the neutral particles can be a little higher than room temperature; however, this temperature rise has not been taken into account throughout this book, unless otherwise stated.

3.2.2.3 Motion in a magnetic field

A particle with mass m and a charge q moves in a magnetic field $\mathbf{B}$ at a velocity of $\mathbf{v}$. A Lorentz force

$$\mathbf{F} = q\mathbf{v} \times \mathbf{B} \tag{3.12}$$

acts on this particle, and its direction is normal to both the directions of the magnetic field and the velocity. When $|\mathbf{v}|$ is constant, this charged particle circulates. The equation of motion is

$$m\frac{v^2}{r} = qv_\perp B \tag{3.13}$$

Here $v_\perp$ is the tangential velocity. The orbit radius r and angular velocity ω are

$$r = \frac{mv_\perp}{qB} \tag{3.14}$$

$$\omega = \frac{qB}{m} \tag{3.15}$$

This motion is called **cyclotron motion**.

The cyclotron motion is utilized in generating a plasma for micro- and nanofabrication. Ions are heavy and therefore are not affected by the magnetic field in such a plasma.

3.3 Collision of Electrons and Molecules

The gas discharge or ionization is caused by the collision of free electrons to gas molecules moving in a space. The collision causes electrons to tear off from the atoms. The electrons inside an atom are strongly bounded to the nucleus, and a large energy is needed to tear off an electron. Different collision processes are present and will be discussed.

3.3.1 Elastic and Inelastic Collisions

When two particles collide, the momentum must be conserved along the directions of the action of force. On the other hand, the kinetic energy is not always conserved. The collision that satisfies the conservation of energy is called **elastic collision**, and the collision that does not satisfy the conservation law is called **inelastic collision**. The energy lost during an inelastic collision transfers to the colliding particles. This energy is spent for ionization or other activation processes. The term collision here can be imaged as a usual impact motion of rigid bodies, whereas, in fact, it is the interaction that occurs when two bodies move very close together.

Let us discuss the energy transfer between particles upon a collision. If a particle with mass m moving at a velocity of v_0 collides with another particle with mass M at rest, the motion after the collision is assumed to be the same as shown in Fig. 3.6.[1] From the conservation of momentum,

$$mv_0 = mv \cos\theta_1 + MV \cos\theta_2 \qquad (3.16)$$

$$mv \sin\theta_1 = MV \sin\theta_2 \qquad (3.17)$$

Now, a part of the kinetic energy ΔU (>0) is lost upon this collision and is converted to the internal energy of the particle.[2] $\Delta U = 0$ is an elastic collision and $\Delta U \neq 0$ is an inelastic collision. The

---

[1] We are not calculating the trajectories of the particles here. That is impossible to do from only the information described here. We are examining how energy is conserved when the particles move in this way.

[2] This "internal energy" is not the same as the total kinetic energy of gas molecules that we have studied in Chapter 2; on the contrary, it refers to the energy state of

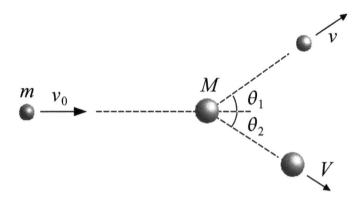

Figure 3.6   Collision between two particles.

conservation equation of energy is

$$\frac{1}{2}mv_0^2 = \frac{1}{2}mv^2 + \frac{1}{2}MV^2 + \Delta U \qquad (3.18)$$

By eliminating θ_1 using Eqs. (3.16) and (3.17), and, furthermore, eliminating v^2 by substituting Eq. (3.18), we obtain

$$\Delta U = -\frac{M}{2m}(m + M)V^2 + (Mv_0 \cos \theta_2)V \qquad (3.19)$$

This is a quadratic function, the graph of which is a convex upward parabola.

The right-hand side of the equation is 0 for the case of an elastic collision ($\Delta U = 0$). Because $V \neq 0$, $V = \dfrac{2m}{m + M}v_0 \cos \theta_2$. By using these relationships, we can derive the ratio of energy transferred from the primary particle (m) to the secondary particle (M), which is called **energy transfer function**.

$$\frac{\frac{1}{2}MV^2}{\frac{1}{2}mv_0^2} = \frac{4mM}{(m + M)^2} \cos^2 \theta_2 \qquad (3.20)$$

---

the interior of an atom or a molecule. When the internal energy of a molecule (or an atom) changes, the electron's orbital changes, leading to ionization or excitation.

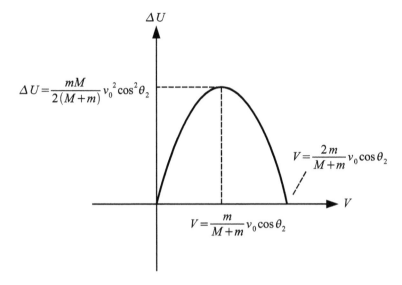

$$\Delta U = \frac{mM}{2(M+m)} v_0{}^2 \cos^2\theta_2$$

$$V = \frac{2m}{M+m} v_0 \cos\theta_2$$

$$V = \frac{m}{M+m} v_0 \cos\theta_2$$

Figure 3.7 Change of internal energy ΔU as a function of V.

From this equation, when the difference in mass is very large, for instance, when $m \ll M$, it is understood that the energy transfer from the primary particle to the secondary particle is negligibly small. It is also understood that this ratio represents the maxima when $M = m$.

We can derive the energy transfer function for an inelastic collision ($\Delta U \neq 0$). The function contains ΔU that is also a function of m and M. Although we cannot determine the value of ΔU uniquely, it is actually not necessary to do so. What we want to do is to understand how the energy transfer function behaves as a function of m and M, therefore, we can set a typical (an arbitral) value to ΔU. Now we use the maximum value of ΔU. ΔU has the maximum at $V = \frac{M}{m+M} v_0 \cos\theta_2$ and is

$$\Delta U = \frac{mM}{2(m+M)} v_0{}^2 \cos^2\theta_2 \tag{3.21}$$

Therefore, the maximum ratio of the energy of the primary particle to the secondary particle is

$$\frac{M}{m+M} \cos^2\theta_2 \tag{3.22}$$

and the secondary particle gains a more internal energy when $m \ll M$. It is noted that the ratio of the transfer of kinetic energy is the same as in the case of an elastic collision,

$$\frac{mM}{(m+M)^2} \cos^2 \theta_2 \qquad (3.23)$$

meaning that the kinetic energy of the primary particle does not transfer to the secondary particle as kinetic energy when $m \ll M$.

We can discuss momentum transfer in a similar manner as we understand that the transfer of momentum MV/mv_0 has a minimum when $m \ll M$. This proves an obvious surmise that the motion of a heavier particle does not change by the impact of a lighter particle.[3]

When $m \ll M$, either for elastic or inelastic collision, the kinetic energy of the primary particle does not transfer to the secondary particle, and the motion of the secondary particle does not change. In the case of an inelastic collision, a part of the kinetic energy of the primary particle is converted to the internal energy of the secondary particle and not to the kinetic energy.

> Electrons in a plasma have much higher energy than ions. As electron mass is extremely smaller than ion mass (approximately $1/10^5$), the motion of molecules is not affected by the collision of electrons. In the case of an inelastic collision ($\Delta U \neq 0$), a part of the kinetic energy of the electron is lost to the molecule, thus raising its internal energy.

Which collision, elastic or inelastic, occurs is a probabilistic problem, and cannot be determined from the discussion made here?

---

[3] This is evident, because when $m \gg M$, the motion of the particle M changes without the transfer of kinetic energy.

Coffee Break: Collision Without Collision?

Can you imagine a perfectly elastic collision without a direct contact?

Let us do a trolley experiment. We have two trollies that have springs on one of their ends. We run the trollies toward each other with the springs facing. As the springs come in contact, they first compress and then stretch back, pushing the trollies away from each other. This is a perfectly elastic collision, although the trollies do not touch each other directly. The same non-contacting elastic collision occurs when magnets are used instead of the springs, with the same poles of the magnets facing each other. When the trollies collide, the kinetic energies of the trollies are stored as the potential energy of the springs or magnets.

If the trollies are pranged, the collision is not perfectly elastic anymore. A part of the kinetic energy of moving trollies is used to break/deform the trollies. Therefore, the conservation of the kinetic energy does not hold. This is an inelastic collision. The motion momentum always holds for moving object unless external forces are subjected.

Electrons travelling toward a molecule see an electron cloud that surrounds the core of the molecule. When the electrons come close to the electron cloud, a similar collision takes place as in the case of the trollies. If the status or structure of the electron cloud changes, that collision is called an inelastic collision.

3.3.2 Collision Processes

Many collision processes are known to occur. The probability of each collision process depends on the kind of atom or molecule and the electron energy. In this section, we study the following six basic collision process. A refers to an atom and XY refers to a molecule.

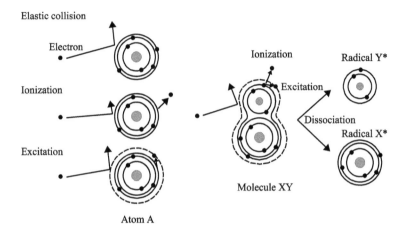

Figure 3.8   Schematics of collision processes.

Elastic collision:	$A + e \rightarrow A + e$
Ionization:	$A + e \rightarrow A^+ + 2e$
Excitation:	$A + e \rightarrow A^* + e$
Relaxation:	$A^* \rightarrow A + h\nu$
Recombination:	$A^+ + e \rightarrow A$
Dissociation:	$XY + e \rightarrow X^* + Y^* + e$

Dissociation occurs only in molecules therefore the symbol XY is used in the equation, whereas for the other processes, it is denoted by the symbol A, to show that it can also occur in molecules. Figure 3.8 illustrates these important collision processes.

Elastic collision   It has already been discussed. In this book, this process is expressed as below.

$$A + e \rightarrow A + e \tag{3.24}$$

$$XY + e \rightarrow XY + e \tag{3.25}$$

Ionization

$$A + e \rightarrow A^+ + 2e \tag{3.26}$$

$$XY + e \rightarrow XY^+ + 2e \tag{3.27}$$

Table 3.3 First ionization potentials of atoms and molecules (eV)

H	13.598	H_2	15.427
He	24.586	N_2	15.576
N	14.534	O_2	12.063
O	13.618	F_2	15.7
F	17.423	Cl_2	11.48
Cl	12.967	CO_2	13.769
Ar	15.759	CH_4	12.704

When a high-energy free electron collides with an atom, an orbital electron—usually an outermost shell electron—frees itself from the bondage of the atom and is released to outside of the atom. This process is called ionization.

The first ionization potential is the energy required to generate a monovalent ion from a neutral atom. The first ionization potentials for some atoms and molecules are listed in Table 3.3.

To maintain a discharge or plasma, ionization must take place continuously; therefore, the ionization is the most important collision process in terms of plasma physics.

Excitation and relaxation

$$A + e \rightarrow A^* + e \qquad (3.28)$$

$$XY + e \rightarrow XY^* + e \qquad (3.29)$$

An orbital electron can transfer to an outer vacant orbital. As the electron in an outer shell has a higher potential energy,[4] this collision process—called excitation—requires energy transfer from the primary electron to the atom.

Symbol * shows that the atom is in the excited state. Unlike ionization, all electrons are bound to the nucleus. The **ground state** is the lowest energy state of an atom and all the orbital electrons are at the lowest position.

The excited atoms are very unstable and, therefore, their electronic configuration reverts to that of the ground state immediately

---

[4] An object under an attractive force field gains larger potential energy as the distance increases. If you raise an object higher, the distance between the object and the center of the earth increases, and the object gain more potential energy due to the gravitational energy.

(in the time range of ns–s). This process is the reverse process of the excitation and is called **relaxation**.

$$A^* \rightarrow A + h\nu \tag{3.30}$$

$$XY^* \rightarrow XY + h\nu \tag{3.31}$$

Here ν is the frequency of the emitted light. The excess energy is released as photon energy. This is why a plasma glows. In this book, relaxation is treated as a collision process.

Recombination

$$A^+ + e \rightarrow A \tag{3.32}$$

Recombination is the reverse process of ionization, where a free electron combines with an ion and the electronic configuration of the atom reverts to that of the ground state. Energy is emitted in an excitation process, and therefore, $\Delta U < 0$. The right-hand side of Eq. (3.21) is always positive. Therefore, the conservation of kinetic energy and motion momentum does not hold for $\Delta U < 0$, and the collision model shown in Fig. 3.6 cannot explain the recombination. In fact, another substance, such as a reactor wall and another gas molecule collide at the same time, the process of which is called three-body collision, and the conservation rule is satisfied. Energy is emitted in the form of electromagnetic waves (light for instance).

Dissociation

$$XY + e \rightarrow X^* + Y^* + e \tag{3.33}$$

Dissociation is breaking up of a molecule. The increment in the internal energy upon inelastic collision is used to break the chemical bonds. Fragments of the dissociated molecule are called **radicals** or free radicals. The radicals have unbounded bonds or lone-pair electrons and show very high reactivity with other molecules. Ar and other mono-atoms do not generate such radicals. The radical concentration of a typical plasma is about 1% and the ion concentration is below 0.01%. Ionized radicals can be formed.

In addition to the collision processes described so far, various other collision processes, such as the generation of negative ions;

collision between neutral species, ions, and excited species; and the involvement of metastable species, are known.

3.3.3 Collision Cross Section

Which type of collision has occurred can be determined stochastically. The collision probability is proportional to the area of interaction between an electron and a molecule. This means that the collision probability can be estimated on the basis of the cross section of the colliding particle. In an extreme case, when the collision should definitely happen, the collision probability is 1 and the collision target size or collision cross section is infinity. In another extreme case, if nothing happens, although the collision occurs physically, the corrosion cross section is zero.

The idea of the mean free path and the collision probability that we introduced in Section 2.6 is now being extended to discuss the collision between an electron and a molecule. That is, as the free electrons in a plasma move extremely faster than molecules, the molecules can be presumed to be at rest. Equation (2.28) is re-written as

$$n\sigma \lambda_e = 1 \tag{3.34}$$

after replacing πd^2 as σ. Subscript e indicates that λ_e is the mean free path of electrons. From Eq. (3.35), collision frequency, θ, is

$$\theta = \frac{\overline{v_e}}{\lambda_e} = \overline{v_e}\sigma n \tag{3.35}$$

confirming that it is proportional to the collision cross section and the molecular number density n.

The collision cross section differs from collision to collision (elastic, ionization, excitation, dissociation) and is dependent on the kinetic energy of electrons. Generally, the cross section of the elastic collision is largest, followed by the ionization cross section, and the excitation cross section. For the case of Ar, the ionization cross section is roughly $1/5$ of the elastic collision cross section, and the excitation cross section is approximately $1/3$ of the ionization cross section. The total of all the cross sections of concern is the total collision cross section (Fig. 3.9).

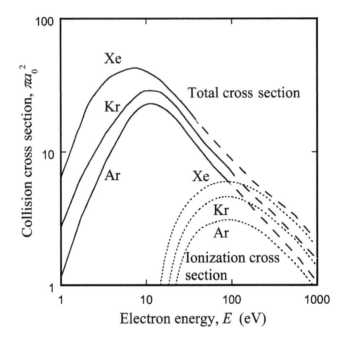

Figure 3.9 Total cross sections of gases. The unit of the vertical axis is πa_0^2. a_0 is the first Bohr radius (0.53 Å).

3.4 Plasma Adjacent to Electrodes

When a plasma is employed for micro- and nanoprocessing, workpieces are placed inside or close to a plasma. Electrodes used for generating plasma and the wall of a vacuum chamber are also present adjacent to the plasma. The plasma itself is neutral, where the concentration of ions and electrons are equalized, whereas its balance is broken at the portions adjacent to an object. The collapse of the balance affects the motion of ions and electrons.

Here we introduce quantities that express the motion of ions and electrons travelling in a particular direction. For this purpose, we use the intensity of an electric current. A current is identical to the amount of total electric charge flowing per unit time. By setting the special density of charged particles to n the velocity of the particles to v the electric charge to q, the electric current per unit or the current density, J, is

$$J = nqv \tag{3.36}$$

The current of ion flow is called **ionic current** and the current of electron flow is called **electronic current**. The ionic current flows in a direction same as that of ions, whereas the electronic current flows in a direction opposite to the flow of electrons.[5] As electrons move much faster than ions (Table 3.1), electronic current is larger than ionic current.

When an object is placed inside a plasma or a plasma is generated close to an object, ions and electrons run out from the plasma toward the surface of the object.[6] As electronic current is much higher than ionic current, the electron density in the region neighboring the object becomes sparse, resulting in the collapse of the neutral condition of the plasma. The particles present in this region are mainly ions and therefore this region is positively charged. Such a charged region is called space charge region. The positive charge generates a positive electric field along the space and increases the voltage potential. The surface of the object is negatively charged due to the presence of the surface electrons,[7] therefore, the electric field points from the object toward the plasma. The plasma is neutral and thus is equipotential over the plasma. Therefore, the electric potential of the plasma is higher than that of the object (Fig. 3.10).

The plasma potential is defined as the electric potential of the plasma against the ground (earth) potential.[8] The space charge region is called **sheath**.

The electric potential at the object's surface is lower than that of the plasma. The electrons are repelled and the electronic current decreases. Contrarily, the ions are attracted by the electrode, the ionic current increases, and the electronic and ionic currents are

---

[5] Recall that we will discuss only positive ions in this book.

[6] The incident electrons will reside at the surface.

[7] The negative charges of the surface attract positive ions. The ionic current and the electric current is finally balanced, but this does not mean that the surface negative charge disappears.

[8] The ground is a portion where electric charge does not accumulate or escapes at ∞ speed ideally. The electric potential of the ground is defined as 0. In actuality, the ground is a large metal component such as casings, chassis, and frame racks that are used the common electrode of electric circuits, the electric potential of which is defined 0 as a reference voltage.

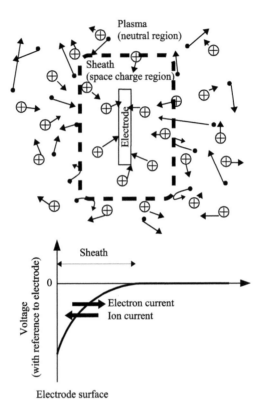

Figure 3.10 Ions and electrons around an object adjacent to the plasma. A sheath is formed around the plasma (electrode). Electrons move at a very high speed and do not go deep into the sheath, which keeps the electrode negatively charged.

equalized. Therefore, it appears that no current flows across the sheath, which means that the sheath behaves as an insulator.

The ion density in this region is almost constant. In such a case, the distribution of electric potential can be obtained by solving a Poisson's equation

$$\varepsilon_0 \frac{d^2 V}{dx^2} = -\rho \tag{3.37}$$

where ε_0 is the permittivity of free space and $\rho(>0)$ is the charge density. The solution of this equation shows a concave downward parabola (Problem 8).

The sheath potential varies, depending on the electric potential of the object.

When the electric potential of the object is not fixed, or "floating" electrically, the electrons incident on the object cannot escape from the object and therefore accumulate on its surface, and the electric potential of the object become negatively charged (taking the ground potential as 0 V), the potential of which is called floating potential. On the other hand, as stated above, the positive charges are piled up in front of the object and the electric potential of the plasma become higher than that of the object.

When the electric potential of the object is fixed, the flux of the charged particles incident on the object varies with the electric voltage applied to the object. For our later convenience, we restate an object as an **electrode**. Not only electrodes for generating a plasma, but objects placed on an electrode and metal vacuum chambers are also regarded as an electrode.

(i) **When the electrode is grounded** $(V = 0\,\text{V})$, the electrons incident to the electrode flow out to the ground. As the positive space charge in front of the electrode raises the potential of the plasma, the plasma potential is positive with respect to the ground, that is, $V = 0\,\text{V}$. Negative charge of an amount same as that of the total amount of positive charge is induced at the electrode surface,[9] therefore, the difference in the potential between the electrode and the plasma is practically identical to the case of the floating electrode.

(ii) **When the electrode is positively biased** against the plasma, by using an external power (voltage) source, the potential difference between the electrode and the plasma is cancelled. The positively biased electrode attracts and extracts the electrons from the plasma, resulting in a large electronic current from the electrode to the plasma.

(iii) **When the electrode is negatively biased**, but the bias voltage is lower than the plasma potential, by using an external power source, the electrons from the plasma are repelled and the ions are attracted to the electrode instead. As ions are very heavy,

---

[9]We can say this is the image charge.

they are not accelerated as magnificently [see Eq. (3.10)], and the final velocity becomes lower than the electrons. Therefore, the ionic current is not as high as the above-mentioned electronic current.

Summarizing the above discussion,

> The ions in the plasma are incident on the surface of an object (electrode) when the electric potential of the object is low. **Sheath** is a region where such a voltage drop exists. Unlike the plasma, it does not glow. The electric potential of the object can be controlled by an external power source.

3.5 Plasma Apparatus and the Interior of a Plasma

3.5.1 DC Glow Discharge

Various discharges are plasmas and the generation methods and causes are different. All the plasmas used for micro- and nanofabrication are generated by applying an electric or electromagnetic field to a rarefied gas. The simplest plasma process is to use a **DC glow discharge**. In practical processes, DC glow discharges are not directly used, and other types of generation methods are employed and/or additional apparatuses are added. Even so, the DC glow discharge is necessarily to be learned, as it is a basic to understand the generation principle and interior structures of plasmas, most of which are common with other types of industrial plasmas.[10]

3.5.1.1 Plasma apparatus

A DC glow discharge plasma apparatus has two discharge electrodes that face each other in a vacuum chamber. The chamber is filled with a rarefied gas, and a high-voltage source is connected to the

---

[10]This book describes plasmas used practically for micro- and nanoprocessing and do not limit to so-called "normal glow discharge."

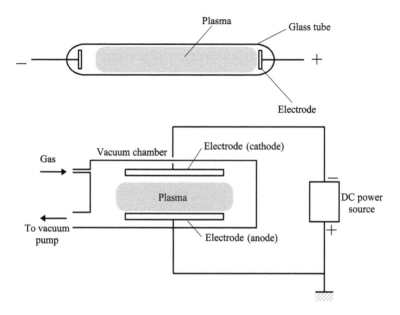

Figure 3.11 DC glow discharge apparatus.

electrodes. The voltage of the anode is 0 V and the voltage of the cathode is approximately 1 kV typically.

Figure 3.11 shows two types of plasma apparatus schematically. In the case of the above schematic, a glass tube filled with a rarefied gas and having an electrode at both ends is used. Neon tubes and Geissler tubes have this configuration. Although this type of discharge tubes are not used for micro- and nanoprocessing, we start our discussion with this most basic discharge apparatus.

Another type of DC glow discharge shown in Fig. 3.11 is a typical plasma apparatus used in practice. The anode and cathode are placed in a vacuum chamber and are connected to a high-voltage source. Piping and flowmeters/controllers to supply a rarefied process gas are attached along with vacuum piping and pumps, and the chamber pressure is maintained at a constant value.

From these schematics, we can notice two main features of a DC plasma discharge apparatus. One is to use a DC (direct current) power source, and another is that the electrodes are encased in a vacuum chamber, or the electrodes are directly exposed to the gas. The generated plasma is slightly apart from the electrodes.

3.5.1.2 Initiation of discharge

After filling a gas to a vacuum chamber at a pressure of, let's say, 10^2 Pa, a voltage is applied to the opposing electrodes. As gases are nonconductive, no current flows to the chamber gas at first. As the applied voltage is increased, a **gas breakdown** occurs at a particular voltage and the current starts to flow. From Fig. 3.12, the breakdown voltage is about 350 V for Ar at 10^2 Pa when the electrode spacing is 1 cm. The initial discharge is arc-like[11] and a lot of ions and electrons are generated. The generated electrons are accelerated toward the anode and gain kinetic energy.

When a highly energetic electron collides with a molecule, two electrons appear, one is the primary electron and the other is from ionization. These electrons get accelerated and collide with other molecules, which multiplies the electrons, leading to an "electron avalanche" or the so-called Townsend ionization.

To start a plasma discharge, the first ionization, or an electron multiplication process, is necessary whatsoever. Natural high-energy sources such as ultraviolet light, naturally occurring radioactive materials, and cosmic rays are thought to induce the ionization that starts a discharge.

3.5.1.3 Structure of DC glow discharge and plasma sustenance

As ions are electrically positive and electrons are negative, they annihilate each other through the recombination. In order to sustain a plasma, high-energy electrons that induce ionization must be supplied from the exterior of the plasma. Its mechanism is related to the behaviors of ions and electrons inside the plasma or in the vicinity of the electrodes. The ions and electrons in a DC plasma are driven and accelerated by an electric field. For this reason, we first

---

[11]An arc discharge is a mode of gas discharge characterized by a high degree of discharge and high current.

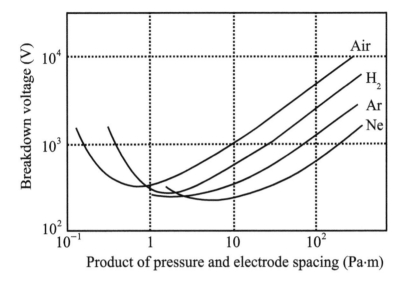

Figure 3.12 Breakdown voltage of various gases (Paschen curve).

study electric aspects of a DC plasma and then the mechanism of electron emission.

Voltage potential of anode and cathode The anode is usually electrically grounded so that its voltage is fixed at 0 V. The potential of the plasma is slightly higher (positive) against the anode.

The voltage potential of the cathode is fixed at a high negative value, and the sheath having a steep potential slope is formed in front of the cathode. The ions in the plasma migrate at a relatively low velocity, drifted by a small voltage slope with random (thermal) motion component, and once they extract or face the sheath, the ions fall the slope and finally collide with the cathode, gaining a large kinetic energy. The impact of ion collision is strong enough to strip the valence electrons off the atoms that consists of the cathode, and these electrons called **secondary electrons** are emitted from the cathode. The secondary electrons travel toward the plasma, in a direction opposite to that of primary electrons, and gain large amount of energy before entering the plasma.

Glow A glowing region is recognized visibly as the body of a DC plasma. Continuous light emission evidences that ion–electron recombination occurs constantly and numerously in the plasma. The high- energy electrons arriving to the plasma from the cathode sheath are deeply involved in the ionization.

In the case of Ar, the energy necessary to invoke the ionization is about 16 eV (see Table 3.3) and electrons having energy roughly higher than this value ionize neutral Ar molecules. The ionization collision cross section is dependent on the energy of primary electrons and becomes smaller for electrons having very high energy (Fig. 3.9).[12] This means that the high-energy electrons coming in from the cathode sheath do not directly contribute to the ionization. These electrons lose their high energy after several collisions in the plasma. When their energies moderate, the collision cross sections of ionization and excitation increase for the collision process to occur.

Sheath The sheaths near the electrodes at both ends do not glow, and this space is called dark space.[13]

The reasons for non-glowing are slightly different for each electrode. For the anode sheath, the energy of the electrons is too low to induce ionization.[14] The cathode sheath potential is high, and the electrons in the cathode sheath have a high energy; the majority have too high energy, as stated above, but there are still a large ratio of electrons that have lower energy that is sufficiently high to cause ionization. However, as the cathode electrons gain high energy, or high velocity, a decrease in the electron density results. (Please recall the definition of flux, or continuity law. As the rate of electron emission at the cathode surface is fixed, the increase in the velocity leads to the decrease in the density.) Therefore, the glow intensity decreases.

The existence of the cathode sheath is extremely important in terms of micro- and nanofabrication. A very intensive electric field in this region accelerates the ions from the plasma and the

---

[12] When a high-speed electron flies near a nucleus, the time during which coulomb interaction works becomes shorter. The momentum of the electron is not altered magnificently. Therefore, the change of the electron velocity as well as the energy loss is small.

[13] The sheath and the dark space are identical in the system of our concern.

[14] The potential of a typical DC plasma is 5–10 eV, which is lower than the first ionization potential of plasma gases.

high-speed (high energy) ions collide with the cathode. This high bombarding energy is utilized to process materials such as in etching and film deposition (see Fig. 3.1. The use of plasmas in micro- and nanofabrication is stated in Chapters 4 and 6 in detail.)

The width of the cathode sheath is dependent on pressure. The sheath width increases when the pressure is low, because the mean free path of the electrons increases. The intensity of the electric field is determined by the voltage drop within the sheath and the sheath width. Therefore, ion bombardment energy can be controlled by adjusting the pressure.

Coffee Break: Plasma Features

A plasma involves many mobile charged particles, namely, ions and electrons. Therefore, the plasma is a good electric conductor. This feature is like that of metals. In fact, metals can be modeled as a solid plasma that consists of immobile ions and mobile electrons moving around the ions freely. As a plasma is a good conductor, its electric resistance is low, and the voltage potential within the plasma is almost constant (see Fig. 3.13). Therefore, a plasma can be extended to any length like the electric wires that can be lengthened and bent. Beautiful neon signs are actually glass discharge tubes bent to desired shapes. Have you seen circular fluorescent lamp bulbs?

(An electric discharge in a fluorescent lamp excites mercury vapor, which produces ultraviolet light, in turn, activating a phosphor coated on the inside of the tube and making it glow.)

3.5.2 RF Plasma

3.5.2.1 Principle and setup

A DC glow discharge uses a current flowing directly from the electrodes to the plasma. Therefore, when the electrode is coated with

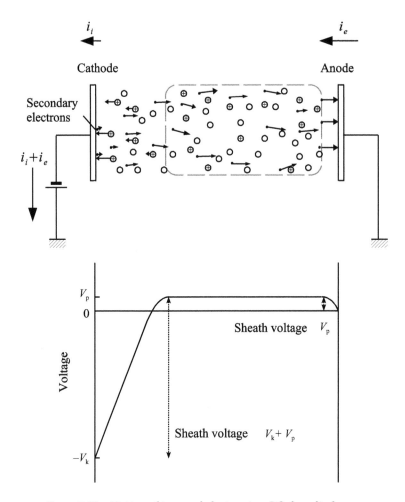

Figure 3.13 Motion of ions and electron in a DC glow discharge.

a non-conductive material or when a non-conductive workpiece is placed to cover the electrode, the discharge does not occur.

This issue is solved by using an AC power source. Plasmas generated by a radio frequency (RF) power source is called RF plasma. Figure 3.14 shows a capacitively coupled plasma apparatus, which is a typical RF plasma generator. It is also called a parallel plate RF plasma apparatus from its configuration.

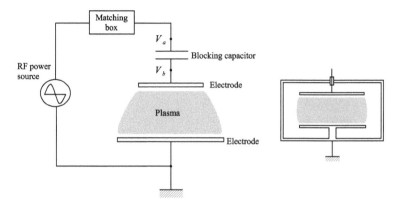

Figure 3.14 Schematics of capacitively coupled RF plasma discharge apparatus.

The plasma discharge is formed between the opposite electrodes, showing a similar configuration as that of a DC plasma apparatus. Two main differences are noted. One is the use of an RF power source and the other is that the plasma does not have to be directly in contact with the electrodes. The latter means that an electrode can be placed outside the vacuum chamber if the chamber (itself or the chamber window for the electrode) is made of a dielectric transparent to RF, such as glass. This system setup is not very commonly used in micro- and nanoprocessing, therefore, metal vacuum chambers will be assumed hereafter.

The vacuum chamber is earthed so that the electric potential of the vacuum chamber is equal to that of the grounded electrode. As an AC is applied, the average voltage of the electrode connected to the power source becomes equal to that connected to the earth. In many cases, a blocking capacitor is inserted between the power source and its electrode. This blocking capacitor isolates or floats a DC component of the voltage applied to the electrode connected to the power source as shown in Fig. 3.14. The role of the blocking capacitor in processing is described later. Another important device but not shown in Fig. 3.14 is a matching circuit that matches the impedance difference between the plasma and the power source. The matching circuit is used to supply the AC electric power to the plasma efficiently.

Figure 3.15 RF plasma discharge apparatus. Adapted from SAMCO web site (http://www.samco.co.jp).

When an AC signal is applied to the electrodes shown in Fig. 3.14, the direction of the electric field alternates to the opposite direction. This is the same as the case that the anode and the cathode of a DC plasma apparatus alternate very quickly. Whether the motion of the ions or electrons can follow such a quick change in the electric direction depends on the rate of the change or the frequency of the AC (RF) signal. At a low frequency, the plasma can follow the change in the direction of the electric field, and the anode sheath is formed at each electrode alternatively, and thus, appears to form at both electrodes in appearance. When the frequency exceeds approximately 10 kHz, ions cannot follow the change in the direction of the electric field because their mass is large (much larger than that of electrons). Contrarily, the electron mass is small and therefore the electrons follow the change and get accelerated by the electric field and gain high energy to reach each electrode. As a result, both the electrodes are negatively charged accompanying a sheath. In either case, a sheath is formed at both electrodes.

The quick alternation of the electric field with a short cycle increases the collision probability between secondary electrons and gas molecules within/near the sheath, which helps to sustain a plasma discharge. Therefore, the RF discharge can be generated in pressures below 1–2 Pa. The discharge occurs easily and the ion current becomes larger at higher frequencies. A very commonly used RF frequency is 13.56 MHz that is specially allocated to industrial equipment worldwide, such as for an RF heating apparatus.

3.5.2.2 Self-bias and its applications

Another very important reason for using an RF plasma is that it enables to use **self-bias**. Self-bias is an offset of a DC voltage superimposed to the RF signal that is applied to the electrode, when a blocking capacitor is inserted between the electrode and the RF power source. In this section, the mechanism of the generation of self-bias is described.

The electrode potential oscillates between positive and negative. When the electrode potential is positive, an electronic current flows to an electrode; and when the electrode potential is negative, an ionic current flows to an electrode. The electronic current across the sheath is always higher than the ionic current, even if the absolute value of the electrode potential is the same. That is, the net current across the sheath flows from the electrode to the plasma, apparently unidirectionally. In this sense the sheath functions as a diode.

To examine the electrical role of this sheath, let us model the blocking capacitor and the sheath of the cathode—the electrode connected to the power source—as an equivalent circuit shown in Fig. 3.16(a). Capacitors are conductive when connected to an AC source as the charge and discharge of the capacitor is repeated cyclically. A diode is a device that shows a low electric resistance when forward-biased ($V_b > 0$) and a high electric resistance when reverse-biased ($V_b < 0$). Therefore, the voltage drop of the diode or the electrode voltage V_b is depicted as shown in the right of Fig. 3.16. The average voltage of this signal, V_{DC}, is negative, which indicates that V_b is the composite signal of the original AC signal, V_a, and a negative DC, $V_{DC}-V_a$ is biased by V_{DC}. This V_{DC} is the self-bias. If the blocking capacitor is not used, the self-bias does not appear and the

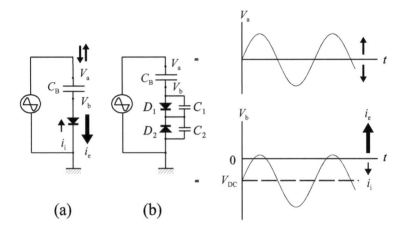

Figure 3.16 Generation mechanism of self-bias.

voltage potentials of the cathode and the anode become symmetric signals.

Then, what happens to the other grounded electrode, anode? Obviously, a sheath exists between the anode and the plasma, which means that our plasma system is modeled as two diodes, the opposite electrodes of which are serially connected. The circuit is symmetric and the current flows to either direction; as far as we use this equivalent circuit, the self-bias is not generated. That is, we notice the existence of a mechanism that suppresses the function of the diode of the grounded electrode.

As described in Section 3.4, a current does not appear to flow until an external voltage potential is applied to the electrode. The sheath is intrinsically an insulator and behaves as a capacitor sandwiched between good conductors, or the electrode and the plasma. We can draw the equivalent circuit including the sheath capacitors as Fig. 3.16(b).

Let us denote the capacitor of the cathode sheath as C_1 and that of the anode sheath as C_2. To come to the point, the self-bias appears when $C_1 < C_2$.[15] The sheath capacitance is proportional to

---

[15]When the current shown in Fig. 3.16(b) flows clockwise, diode D_1 is turned on (in the conduction mode) so that C_1 does not store electric charge. This circuit is simplified so that C_B and C_2 are connected in series. Likewise, for the

the area of the electrode, so that this condition is realized if the area of the grounded electrode (anode) is larger than that of the power-supplied electrode (cathode). In actual apparatuses, this condition is fulfilled, as the metal vacuum chamber is used as a common electrode and grounded. The anode is electrically connected to the chamber so that the area of the anode is relatively small (see Fig. 3.14).

An RF plasma with self-bias possesses advantages of both the DC plasma and the RF plasma. The electrode connected to the power source is negatively DC biased and functions similar to the cathode of the DC plasma, which enables to extract the ions from the plasma. A workpiece is processed by the irradiation of the extracted ions. In addition to this, the plasma can be sustained even if an insulative workpiece is placed on the electrode, which enables to process insulative materials.

In general, the ions in the plasma gain higher energy as the frequency of the power source increases, similar to the forced oscillation of a mass point. This also makes the RF plasma more attractive and useful than the DC plasma.

3.5.3 Development of Plasma for Micro- and Nanofabrication

It is advantageous to increase the ratio of ions and radicals in a plasma for micro-/nanoprocessing. For precise fabrication, it is important to decrease the operating pressure so as to increase the mean free path of particles and increase the reach of the linear motion of ions. However, the density of ions (as well as radicals) decreases as the pressure decreases as long as the ionization rate is unchanged. Therefore, high-density plasmas that provide a higher ionization ratio are preferred in advanced processes.

In this section, an inductively coupled plasma (ICP) and a magnetically enhanced plasma are described as representatives of plasmas used for advanced micro- and nanofabrication.

---

counterclockwise current, C_B and C_1 are in series. When $C_1 < C_2$, the voltage of element C_1 becomes small for the clockwise current, and the voltage of element C_2 becomes large for the counterclockwise current, where D_1 functions more effectively than D_2.

3.5.3.1 Inductively coupled plasma

An **inductively coupled plasma (ICP)** can be generated by applying an RF power to a coil at the axial center of which a quartz tube is placed. The quartz tube is obviously evacuated (vacuum). In the cases of plasmas discussed so far, the motion of the electrons are driven by an electric field. In contrast, the ICP is sustained by electrons driven by the change of the magnetic field. The ICP provides a higher plasma density (10^{11} cm^{-3}) than capacitively coupled plasmas, because the coil generates a concentrated magnetic field easily and the Lorentz motion of electrons assists in ionizing molecules.

This simple configuration using a quartz tube was not very popularly used for advanced industrial apparatuses. The plasma is formed very closely to a quartz tube wall, leading to the sputtering of the tube, which contaminates the plasma. Recent apparatuses utilize more sophisticated methods to couple the plasma and the RF power source, for instance, using an "antenna" that emits the RF power inside the metal vacuum chamber (Fig. 3.17). The configuration of ICP is simpler (at least than the ECR plasma described next) and therefore the ICP is widely used in micro-/nanoprocessing.

3.5.3.2 Magnetic field and ECR plasma

Here, the generation of a magnetically enhanced plasma and its application to **electron cyclotron resonance plasma** are briefly described.

As stated in Section 3.2.2.3 (p. 64), when the plasma is subjected to an external magnetic field, the electrons have an additional velocity components normal to both the direction of the magnetic field and that of the motion due to the Lorentz force $\mathbf{F} = q\mathbf{v} \times \mathbf{B}$. If $|\mathbf{v}|$ is constant, the electron in such motion have a circular trajectory with a radius of $r = mv/qB$. The ions in the plasma behave the same, whereas their motion is not susceptible to this effect because the ions are very heavy.

When the electrons circulate, their travel distance increases enormously. The higher the travel distance, the higher the rate of ionization or plasma density. In addition to this effect, a significant

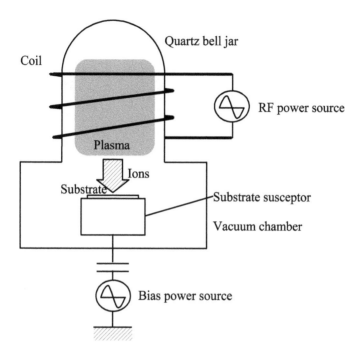

Figure 3.17 Inductively coupled plasma apparatus.

increase in the ionization rate can be achieved by using a specific configuration. The **magnetron plasma** is the representative, in which the circularly traveling electrons are arranged to collide with the cathode that emits the secondary electrons. We will study the magnetron plasma later in more detail in Chapter 4.

The periodic (angular frequency) of rotation is $\omega_0 = qB/m_e$ [Eq. (3.15)]. In an AC electric field, which has the same frequency as ω_0, the electrons continue to rotate (resonance) so that the plasma density increases significantly. This kind of plasma that is specially arranged to sustain the cyclotron motion of the electrons is called ECR plasma. The ECR condition for a magnetic field of 900 gauss (90 mT) is $\omega_0 = 1.6 \times 10^{10}$ rad/sec or 2.5 GHz. This is a microwave that is introduced to the vacuum chamber through a quartz window from a waveguide (Fig. 3.18).

The ICP plasma and ECR plasma are called an electrodeless discharge because they do not use electrodes in literal sense. The

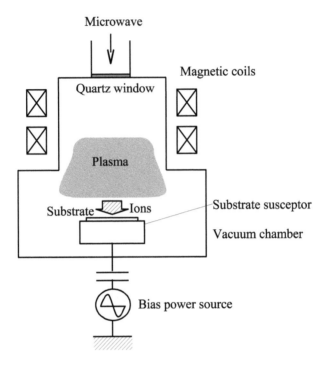

Figure 3.18 ECR plasma apparatus.

electrodeless discharge has an advantage that the plasma potential and the substrate (workpiece) potential can be controlled independently. The plasma potential is positive ($V_p > 0$) with reference to the grounded chamber). If an electrode biased negatively $V (< 0)$ is placed adjacent to or away from the plasma, the ions are extracted from the plasma toward the electrode. The energy of the ions, qV, is determined independent of the condition for plasma generation, such as the electric power. Either a DC or an RF power source can be used for biasing the electrode. The DC biasing is simpler but the RF biasing is more convenient and popular as either conductive or insulative substrate/workpiece can be used.

The conditions of a plasma, such as plasma density and sheath potential, are largely influenced by the potential of the electrode or the power of RF source. For example, when the RF power is raised, the plasma density increases, whereas the self-bias V_{DC} also increases resulting in an increase in the ion-bombardment energy or the incident-ion flux. Too intensive ion bombardment is not always preferred due to concerns of over processing and material damage.

In advanced micro- and nanoprocesses, lower pressure and higher plasma density are demanded so as to reduce the plasma damages. A combination of the electrodeless discharge and the independent substrate bias has become an essential technique.

Problems

(1) A particle with a mass m and a charge q (<0) at rest is accelerated by the voltage potential difference V (>0). What is its final velocity v?

(2) Describe the reason why electrons in a plasma have a larger kinetic energy than ions. Use equations if necessary.

(3) Explain the cyclotron motion of an electron.

(4) Derive Eq. (3.19).

(5) Derive energy transfer function [Eq. (3.20)]. Then examine the behavior when $m \ll M$.

(6) List at least four collision processes and explain each of them.

(7) Calculate the ionic and electronic currents, using the values listed in Table 3.1.

(8) Plot the electric potential distribution for $x \geq 0$ by solving the Poisson Eq. (3.37) under the boundary conditions of $V = 0$ at $x = 0$ and $dV/dx = 0$ for $x \geq \ell$.

(9) Draw a schematic diagram of the DC glow plasma apparatus and explain how the plasma is sustained.

(10) Explain the role of a blocking capacitor of an RF plasma apparatus from both the electrical circuit and the micro-/nanoprocessing point of views.

(11) Describe the features of an electrodeless discharge.

Chapter 4

Physical Vapor Deposition

4.1 Introduction

A **thin film** is a thin layer that has thickness less than 1 μm or less than a few μm at the thickest. Thin films are formed on a **substrate** that is a thick (compared to the film) rigid plate or wafer, because delaminated films may have poor mechanical properties, and moreover, they are too difficult to handle.[1]

In this book, solid thin films are considered as the material used to fabricate very small components for electronic devices. For instance, to fabricate a wire that has a cross section of 1 μm × 1 μm, a 1 μm thick film is deposited and then patterned to 1 μm width (see Chapter 6). Thin-film deposition is an additive process in which the film is added to a substrate. Thin films are widely used in many fields as high-performance materials, such as for large-scale integrated circuits, optical coatings, and magnetic discs and heads.

In this chapter, **physical vapor deposition**, abbreviated as **PVD**, is discussed as a representative and most common technique for thin film deposition in micro- and nanoprocessing. PVD is a technique to form a thin film by laying atoms and/or ions that are produced

---

[1]A thin film of Cu is called a Cu thin film but not a thin Cu film.

Micro- and Nanofabrication for Beginners
Eiichi Kondoh
Copyright © 2021 Jenny Stanford Publishing Pte. Ltd.
ISBN 978-981-4877-09-1 (Hardcover), 978-1-003-11993-7 (eBook)
www.jennystanford.com

"physically," onto a substrate. Substrate heating is not necessarily used in principle; low temperature–resistant materials can be used as a substrate.

There are a wide variety of PVD techniques and derivatives known. In this chapter, we will study the most fundamental techniques—evaporation and sputtering. The background knowledge needed here is gas kinetics and basics of plasma physics.

4.2 Evaporation

4.2.1 Evaporation and Deposition

4.2.1.1 Vacuum evaporation

When you take a bath or shower, the mirror in the bathroom will fog up immediately, especially in winter. Your bathroom is filled with water vapor, and the water vapor condenses on the cool mirror. If the temperature of the mirror is far below the freezing point of water, an ice film will be formed.

The same thing happens when a metal is vaporized and the vapor is exposed to a substrate. Even if the substrate is heated to several hundred degrees Celsius, a solid film grows on the substrate, as the melting point of metals is high. In general vapor deposition refers to a method to grow a solid thin film on a solid substrate.

A general meaning of **evaporation** is the vaporization of a liquid, but in thin film technologies, evaporation indicates a method to obtain solid thin films by vaporizing a solid under vacuum; and the word evaporation will be used in this sense throughout this book. Evaporation is often referred to as vacuum evaporation, vacuum deposition, thermal evaporation, and even vapor deposition.

4.2.1.2 Vacuum evaporator

As it will be useful to understand later topics, the construction of a vacuum evaporator is shown here. Figure 4.1 shows an example of the simplest setup. A source material, usually pieces of a metal, is placed in a refractory metal boat that is located at the lower end of a vacuum bell jar. The metal boat is electrically heated. As the

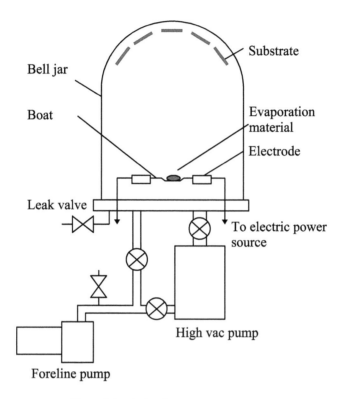

Figure 4.1   A simple vacuum evaporator.

temperature of the source material increases, it starts to release its vapor. The substrates are placed facedown so as to catch the upcoming vapor on their surface.[2] When a film grows to a desired thickness, the heating is stopped and the substrates are retrieved after breaking the vacuum.

The bell jar is evacuated with a high-vacuum pumping system below at least 1×10^{-3} Pa before evaporation (background pressure). The pressure of the bell jar becomes higher during evaporation due to the metal vapor and sometimes reaches as high as 1×10^{-1} Pa.

---

[2]The substrates cannot be placed faceup as the boat needs to support a melt.

4.2.1.3 Vapor pressure

Molecules of a liquid get thermally activated and start vibrating. The distribution of the energy almost obeys the Boltzmann distribution. This means that there are always molecules that have energy sufficiently high enough to desorb (desorption) or to vaporize (vaporization), although the ratio of such molecules can be extremely small. Similarly, molecules existing as a vapor obey the Maxwell–Boltzmann distribution, in which some of the molecules that have small energy can be trapped when they collide with the liquid surface (**condensation**).

Therefore, although the "stable phase" of H_2O in our ambience is liquid, there is always some amount of water vapor present in the air. Ideally, a liquid and its vapor are in equilibrium, and

$$\text{Rate of vaporization} = \text{Rate of condensation} \qquad (4.1)$$

The (partial) pressure of the vapor under this condition is called **equilibrium vapor pressure**. Each material has an equilibrium vapor pressure. We denote the equilibrium vapor pressure by p_e.

Atoms and molecules that form a solid substance are also in thermal motion, not as strongly as liquid, but can leave the solid surface. This is called sublimation. In thin film technologies, we include sublimation with the "evaporation" method. At room temperature, the sublimation of solid metals does not occur because of the too low equilibrium vapor pressure and the contaminants and/or oxides that cap their clean surfaces. Vaporization can occur significantly when the temperature is increased near to the melting point because then the contaminants and oxides decompose and/or vaporize, resulting in cracks in the oxide skin that reveal a clean liquid surface.

The relationship between equilibrium vapor pressure and temperature is expressed by the Clausius–Clapeyron equation

$$\frac{dp_e}{dT} = \frac{\Delta H(T)}{T \Delta V} \qquad (4.2)$$

where T is the absolute temperature (K), $\Delta H(T)$ is the heat of vaporization at T (J/mol), and ΔV is a volume change upon the vaporization (m^3/mol). Usually,

Volume of vapor V_{vapor} > Volume of condensed phase $V_{cond.}$

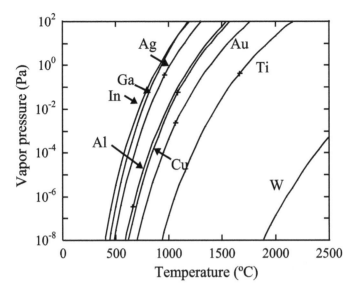

Figure 4.2 Temperature dependence of the equilibrium vapor pressure p_e of metals (+ : melting point).

and therefore, we can assume

$$\Delta V = V_{vapor}$$

In addition, the temperature dependence of the heat of vaporization is not large so that we can assume that it is constant. By substituting the equation of state $pV = NRT$ (here $N = 1$), Eq. (4.2) is reduced to

$$\frac{dp_e}{dT} = \frac{p\Delta H(T)}{RT^2}$$

and after solving this equation for p_e, we obtain,

$$\ln p_e = -\frac{\Delta H(T)}{RT} + \text{constant} \qquad (4.3)$$

This equation shows the temperature dependence of the equilibrium vapor pressure p_e.

Figure 4.2 shows the temperature dependencies of the equilibrium vapor pressure of elements (vapor pressure curves).

The equilibrium vapor pressure increases exponentially with temperature [$p_e \propto \exp(-\Delta H/RT)$]. It should be noted that the equilibrium vapor pressure does not significantly increase just when the temperature exceeds the melting point, as understood from the continuity of the curves. Indeed, when depositing high-melting point metals, such as Cr and Fe, the evaporation is carried out without melting the evaporation sources. However, we usually obtain better vaporization efficiency by melting the source metals.

4.2.1.4 Vaporization rate

Gas molecules travel freely in a space and can collide with a surface incessantly. At a gas pressure of p (Pa), the number of molecules incident on a unit area (here 1 m²) of the surface is

$$\sqrt{\frac{1}{2\pi mkT}}\, p_e \quad \text{(atoms/m}^2\text{/s)} \tag{4.4}$$

where m (kg) is the mass of the molecule [Eq. (2.24)]. In equilibrium, rate of vaporization = rate of condensation, which means that the amount of vaporizing atoms is equal to the amount of atoms returning from the gas phase to the surface of the evaporation surface. Therefore, by using the gas equilibrium pressure of p_e instead of p of Eq. (2.24), we obtain the flux of evaporation; and furthermore, by multiplying the area of evaporation A_{source} to this equation, we obtain the equation of evaporation rate:

$$\sqrt{\frac{1}{2\pi mkT}}\, p_e A_{\text{source}} \quad \text{(atoms/s)} \tag{4.5}$$

By multiplying this equation with the atomic mass, m, we obtain the mass of the source material converting to vapor per second, or the evaporation rate:

$$\sqrt{\frac{m}{2\pi kT}}\, p_e A_{\text{source}} \quad \text{(kg/s)} \tag{4.6}$$

As the chamber background pressure is quite low, the mean free path of the molecules is very large, which means that the molecules leaving the evaporation source travel linearly without colliding much with other gas molecules. Most of the molecules attach to the chamber wall and some of the molecules attach to the substrate, and some are evacuated by the pump; therefore the evaporated atoms do not return to the source.

The kinetic energy of the atoms originates from thermal energy. If the source temperature is 1000 K, the thermal energy is approximately

$$kT = 0.1 \text{ eV}$$

The evaporated molecules retain this energy gained by temperature increase until they collide with the surface of the wall or substrate. The wall is much cooler and the molecules condense on the solid by losing their thermal energy. One might ask about the temperature rise of the substrate. This happens because the energy released from the evaporated atoms transfers to the substrate. However, as the depositing film is thin, and the substrate has a much larger heat capacity and are often very thermally conductive, the rise of the substrate temperature is negligible in the extent of usual evaporation experiments.[3]

4.2.1.5 Deposition rate and film uniformity

The amount of film deposited per unit time is called the **deposition rate**. The commonest unit is thickness change per unit time, for instance nm/s, and the mass change per unit time is also used, such as mg/min.

Taking the film thickness at time t as D, the deposition rate R

$$R = \frac{dD}{dt} \tag{4.7}$$

or

$$D = \int_0^t R dt \tag{4.8}$$

from its definition. In PVD, the deposition rate becomes constant under fixed deposition conditions. Therefore, the thickness increases in proportion to time.

Taking the flux of atoms incident on the substrate as J, the mass depositing per unit time and per unit area material is mJ

$$R = \frac{mJ}{\rho} \tag{4.9}$$

---

[3] However, this is not the case for sputter deposition.

where ρ is the mass density of the depositing material. The units of mJ is $kg/m^2/s$ when we use $atoms/m^2/s$ for J and kg/m^3 for ρ.

Vapor from the evaporation source expands in the chamber and a part of the vapor reaches the substrate. Due to this expansion, the flux of atoms incident on the substrate, J, becomes smaller than the flux of evaporation [Eq. (4.4)]. Its ratio is deposition area divided by the total expanded area. Therefore, the number of atoms incident to an area of the substrate is

$$\text{Number of incident atoms} = J \times \text{Incidence area}$$
$$= \frac{\text{Evaporation flux} \times \text{Evaporation area}}{\text{Total surface area of expansion}} \times \text{Incidence area}$$

$$(4.10)$$

This tells us the relationship between the evaporation rate and the deposition rate.

The sizes of vacuum chambers and substrates are several tens of centimeters to several meters and several centimeters to several tens of centimeters. The evaporation source is much smaller than these and can be considered to be an infinitesimally small point. The expansion from a spherical point source is isotropic or spherical so that the total surface area of expansion is $4\pi r^2$, where r is the distance between the point source and the sphere surface. By setting the surface area of the point source to A_{source} and by using Eq. (4.5) and Eq. (4.10), the atom flux at the sphere's surface, the radius of which is r, is:

$$\sqrt{\frac{1}{2\pi kTm}} \frac{p_e}{4\pi r^2} A_{source} \quad (\text{atoms or molecules/m}^2) \qquad (4.11)$$

When substrates are placed in contact with this hypothetical surface, the incident flux is the same all over the surface or the substrates so that the deposition rate is

$$R = \sqrt{\frac{m}{2\pi kT}} \frac{p_e}{\rho} \frac{1}{4\pi r^2} A_{source} \qquad (\text{m/s}) \qquad (4.12)$$

Actual substrates are not located at points equidistant from the evaporation source, which means that the thickness of the film varies with its position on the substrate. For instance, in Fig. 4.3, a

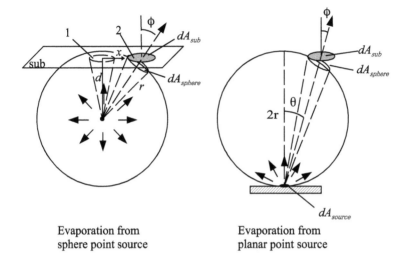

Figure 4.3 Evaporation from spherical and planar points.

thicker film deposits on region 1 on plane "sub" than on region 2. This is because region 2 is away from the evaporation source and tilted at an angle ϕ so that the projection area increases and the incident flux decreases (recall Section 2.5.3 cosine law). Therefore, the area of incidence and the evaporation area in Eq. (4.10) are defined as infinitesimally small.

The area of region 2 dA_{sub} is a projection of the small area on the (expanded) sphere dA_{sub} at the same position as region 2.

$$dA_{sub} = \frac{dA_{shpere}}{\cos\phi} \qquad (4.13)$$

Therefore, the flux of molecules incident on dA_{sub} is $\cos\phi$ times smaller than that on the sphere's surface.

On taking the length of the perpendicular from the point source to the plane "sub" as d and the distance from the foot of the perpendicular to a given point as x, we obtain $r = \sqrt{d^2 + x^2}$ and $\cos\phi = d/\sqrt{d^2 + x^2}$. Therefore, from Eq. (4.12), the deposition rate at the position x—more precisely, at an infinitesimally small region—is:

$$R(x) = \sqrt{\frac{m}{2\pi kT}} \frac{p_e}{\rho} \frac{1}{4\pi} \frac{d}{(d^2 + x^2)^{3/2}} A_{source} \qquad \text{(m/s)} \qquad (4.14)$$

As the thickness is proportional to the deposition rate, the thickness distribution becomes

$$\frac{D(x)}{D(0)} = \frac{R(x)}{R(0)} = \frac{1}{[1 + (x/d)^2]^{3/2}} \qquad (4.15)$$

where the thickness at $x = 0$ is normalized to 1.

When a planar vaporization source is used, the cosine law at the source surface—the areal projection to the source surface plane—is to be considered additionally.

When viewing an aerial element on the planar source dA_{source} from the direction tilted by an angle θ, the area of the element is $dA_{\text{source}} \cos\theta$ so that the molecular flux becomes $\cos\theta$ times smaller than in the case of the point source. When the source plane and the substrate are parallel, $\phi = \theta$, and therefore, the deposition rate and the thickness distributions are

$$R(x) = \sqrt{\frac{m}{2\pi kT}} \frac{p_e}{\rho} \frac{1}{4\pi} \left(\frac{d}{d^2 + x^2}\right)^2 dA_{\text{source}} \qquad \text{(m/s)} \quad (4.16)$$

and

$$\frac{D(x)}{D(0)} = \frac{1}{[1 + (x/d)^2]^2}, \qquad (4.17)$$

respectively.

So far, we have discussed idealized models, but these models can be applied in many practical cases. A molten metal droplet attached to a heating filament, of which details are described later, is modeled as the point evaporation source. An evaporation source, which is a crucible whose lid has a small opening for the release of vapor— called a Knudsen cell—is modeled as a planar point source.

For general configuration of sources and substrates, the geo-metrical relationship is taken into account in the deposition rate equation, which is expressed as the integration

$$\int_{A_{\text{source}}} \int_{A_{\text{sub}}} \frac{\cos\theta \cos\phi}{4\pi r^2} dA_{\text{source}} dA_{\text{sub}} \qquad (4.18)$$

It is not straightforward at all to solve this equation analytically, but there are some known solutions for special cases (see Section 4.3.1.4).

The film thickness profile is obtained by plotting Eq. (4.15) and Eq. (4.17). An example is shown later in Fig. 4.13. It is understood that the film thickness decreases precipitously with the distance from the vaporization source. In order to obtain a uniform film,

- Increase the distance between the source and the substrate (but at the expense of deposition rate lowering)
- Tilt the substrate so that the normal points toward the source
- Change the position of the substrate with time (for instance, substrate rotation)
- Use a planar and large evaporation source (not very common due to technological difficulties)

are listed as solutions.

4.2.1.6 Multicomponent deposition and impurity incorporation

Alloys and compounds consist of different elements. In order to obtain an alloy or a compound film, multiple evaporation sources are used simultaneously.

A conceptual schematic of a multisource evaporation is shown in Fig. 4.4. Generally, for multi-element deposition, an alloy is used as a source and evaporated, or multiple sources are used to evaporate different elements at the same time. If the materials do not react with each other, a film of a solid solution or multi-phase alloy is deposited.

When a solid solution alloy is used as an evaporation source, each component element vaporizes independently and the film of a similar or the same alloy is obtained. When a compound is used, it is not very common for the compound to vaporize itself, instead it generally decomposes into constituent elements and reaches the surface as a single element or as a simple molecule. The deposited elements may form the same compound as the source but it depends on the various deposition conditions. In general, substrate temperature is the most important factor to govern the properties of the film.

When materials A and B reach the substrate surface with incident flux of J_A and J_B, respectively, the average composition of the deposited films is expected to be simply

$$J_A : J_B$$

When a multicomponent (alloy or compound) source is used, the composition of the deposited film may not be the same as that of the

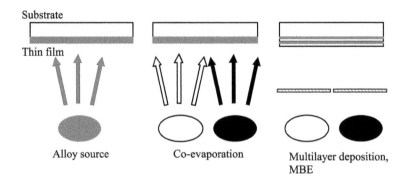

Figure 4.4   Vacuum evaporation of a multicomponent film (conceptual).

source. For an A–B binary evaporation source material, the molar ratio of which is $x_A : x_B$, and the incident flux is given by Eq. (4.4), the film composition becomes

$$\frac{x_A p_A}{\sqrt{M_A}} : \frac{x_B p_B}{\sqrt{M_B}} \tag{4.19}$$

where M_A and M_B are the molar weights and p_A and p_B are the equilibrium vapor pressures.

As the equilibrium vapor pressures differ from one component to another, the composition of the alloy or compound source changes, leading to a change in the concentration of the low-vapor pressure element; thus, resulting in continuous change in the film composition when the time of the evaporation operation is long. Simultaneous deposition using multiple sources—often called co-deposition—is an ideal and practical solution to avert this issue but precise control of deposition conditions (temperature and source area) is required to obtain the desired film composition.

Impurity gases incorporated in the vacuum can be easily picked up by the film during deposition. Especially, O_2 and H_2O are reactive and are a major factor of impurity incorporation. The incident flux of molecules of an impurity species is expressed by Eq. (4.4), substituting p_e with the partial pressure of the impurity gas p_i. The impurity uptake can depend on the combination of the evaporating material and the impurity gas. If the impurity gases are inert or if the evaporating material is non-reactive, the impurity gas may desorb and thus may not be incorporated in the film. In the case of the combination of reactive species such as Al, Ti, and O_2, H_2O, almost all the impurity gases are incorporated (as O atoms in this case) in the film. Interestingly, this is applied to reduce gaseous impurities in the chamber, where such metals are evaporated before the main deposition run.[4]

Figure 4.4 depicts another type of co-deposition other than the deposition of mixture. A shutter is placed at the outlet of each evaporation source, and each shutter is opened one after the other. This technique is used to deposit multilayer films.

Coffee Break: Superlattice and Molecular Beam

A crystal consists of atoms positioned periodically, and the atoms form a lattice. A superlattice is an artificial crystal formed by layering a very thin film of different elements periodically. In 1969, Dr. Leo Esaki of IBM predicted that semiconductor superlattices possess unique electronic properties and proved it in GaAlAs superlattices.

The technique that Dr. Esaki and his colleagues used to realize the superlattice is molecular beam epitaxy. A molecular beam is a linear stream of molecules formed by evaporating the source material in ultra-high vacuum. These molecules travel in a straight line, without colliding with other molecules. The outlet of the beam is carefully designed to tailor the shape and direction of the beam. The use of the molecular beam enabled precisely control the incident flux on the substrate and helped in growing high-quality films by suppressing undesired three-dimensional growth.

It is important to note that Dr. Esaki received a Nobel Prize in physics in 1973 for the discovery of the tunnel effect in a semiconductor pn junction. The invention of the superlattice is as valuable as this.

---

[4]A shutter is positioned between the source and the substrate to prevent undesired deposition.

4.2.2 Evaporation Sources and Derivative Methods

4.2.2.1 Resistive evaporation

Resistive evaporation or resistive thermal evaporation uses a refractory metal (such as W and Ta) resistive heater to melt a source metal. A simple heater is a straight or a basket-like coil (often called "filament"). Boats are also widely used. The refractory metal heaters should be strong enough to not get deformed by self-weight or of the source material at elevated temperatures. A typical diameter of a reaction with the filament wire is 0.5–2 mm (Fig. 4.5).

Usually the source material is directly placed on the heater. When a filament is used, the droplet melt is supported by its surface tension and the wetting between the melt and the filament. The heating current must be controlled carefully. Too high a temperature will result in the melt losing its viscosity and inducing an unfavorable reaction between the melt and the filament, whereas too low a temperature will result in poor vaporization. The use of

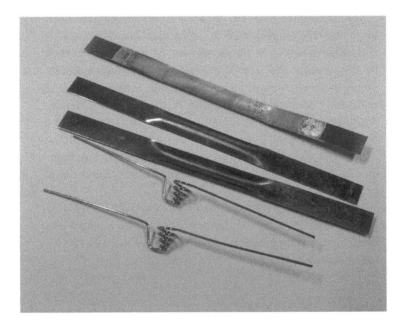

Figure 4.5 Resistive evaporation heaters (top one is after use).

a boat can evade this issue; the penalty is requirement of very high electric power.

Due to their simple arrangement and easy operation, thermal evaporators are widely used from industry to laboratories. Major drawbacks are difficulty in controlling the deposition rate, degassing and impurity from the filament, and rather low heating temperature (usually lower than 1500 K).

When direct contact of the source with the filament is not desired, a ceramic crucible is used along with a basket filament. Obviously, the heating efficiency is deteriorated. The use of inductive heater (furnace) increases the heating efficiency significantly and provides better temperature controllability. The contamination from the crucible (which is made of substances such as BN and Al_2O_3) is a concern.

4.2.2.2 Electron beam evaporation

In electron beam (e-beam) evaporation or deposition, the evaporation source is locally irradiated with an electron beam so as to heat and melt the source material. The source is placed in a water-cooled, copper crucible so that only the central portion of the source material is heated resistively.[5] If the source material is same as the material of the inner crucible, the chances of impurity incorporation from the crucible are eliminated. Local-point heating by an e-beam helps to agitate the melt pool, which improves controllability and reproducibility. Most of the e-beam sources have a thermal electron emitter, which is a heated filament, and a strong magnet collimates and bends the electrons to target them to the center of the evaporation source. The irradiation position of the electron beam can be controlled more precisely by using electromagnetic assemblies, which also allows surface scanning and wider surface heating.

As the principle used for heating is resistive (Joule) heating, the melting of the electron beam allows vaporization of the metals that have a high melting point. The current adjustment makes

---

[5] An electron beam proceeding in a vacuum is an electric current and is identical to an electric wire connected to the source material.

a close influence on the melting, and the temperature and the deposition rate can be controlled much better than in resistive thermal evaporation. For these reasons, electron beam deposition is widely used in industries as well as in laboratories. When depositing alloys or compounds, multiple sources are prepared. In this case, multiple e-beam sources (filaments) are used, but a single source can also be used when the evaporation sources are heated one-by-one. It should be noted that an electron beam irradiation can induce emission of characteristic X-rays from the source metal if a high-voltage electric source is used. In such a case, a metal chamber must be used and the deposition of films that requires high electronic reliability, such as for electronic devices, must be carried out carefully.

4.2.2.3 Reactive evaporation

Reactive evaporation is a method to carry out evaporation by adding a small amount of a reactive gas. The vapor and the gas react with each other, usually on the surface of the depositing film. It can be called a chemical physical vapor deposition.

4.2.2.4 Ion plating

Ion plating is a method to deposit a film by ionizing the vapor. The ionized vapor is accelerated and extracted by an electric field. It has the advantage of evaporation as well as that of sputtering or ion beam deposition. The acceleration increases the energy of ions that allows to form a dense high-quality film at a high deposition rate (see also Section 4.3). The principle of using an electric field to extract ions is similar to electroplating, and this method is named ion plating, and is expected to replace dry electroplating in which a large amount of toxic liquid waste is produced.

An ion plating apparatus has a device that ionizes the vapor and accelerates the ions in addition to an evaporation source. An inert gas plasma or electron beam is used for ionization of the vapor. Typical operating pressures are $10^{-2}-10^{0}$ Pa, one or two orders of magnitude higher than those in usual evaporation. This

Figure 4.6   Electron beam evaporation source with 3 crucibles and 1 e-gun.

helps in producing particles having various physical states through collisions with other molecules, ionization, and neutralization. The ionization ratio is about a few percent. The bias applied between the plasma and the substrate is several tens to several hundreds volts. Too high a bias induces a decrease in the deposition rate due to sputtering.

Various compounds, such as metal oxides, carbides, and nitrides can be deposited by admitting a reactive gas. Ion plating is used to coat materials to which electroplating is hardly applied and to deposit dense and thick films.

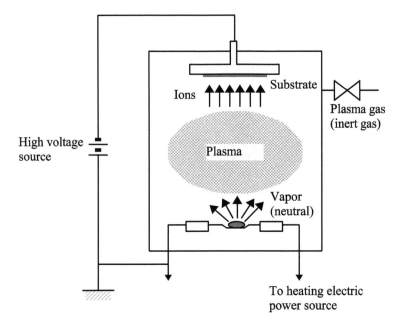

Figure 4.7   Ion plating.

4.2.3 Features of Vacuum Evaporation

Before closing this section, the features of vacuum evaporation are listed.

  (i) Simple configuration of apparatuses especially thermal evaporation; alloys and compounds having a wide variety of composition can be deposited by using multiple sources.

 (ii) Electron beam deposition has superior deposition rate controllability and produces highly pure films.

(iii) Atoms incident to the surface have a small kinetic energy (thermal energy), which suppresses the increase of the substrate temperature.

(iv) Particles incident on the substrates are neutral species. In semiconductor technologies, device damages due to energetic charged particles from a plasma are a concern, whereas in thermal evaporation there is no such danger.

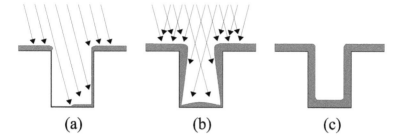

Figure 4.8   Step coverage by a film.

(v) **Step coverage** (or **film conformability**) is poor.[6] As atoms
come straight to the surface like a ray, a thinner film deposits at
the part shadowed by a step. Contrarily, a thicker film deposits
at a sharp corner because this part has a larger solid angle.

4.3 Sputtering

4.3.1 Principle of Sputtering

Sputter deposition or sputtering is another PVD method. A plasma
is generated using gases, such as Ar, and ions in the plasma are
extracted and irradiated to an ingot of source material, so called the
target. The Ar ions sputter (described later) the target particles that
are incident on the substrate placed opposite to the target. Sputter
deposition is a representative PVD method comparable to vacuum
evaporation and is widely used for the deposition of various metals,
alloys, semiconductors, and compounds.

Figure 4.9 shows an example of sputter deposition apparatus.
The target is located opposite to the substrate pedestal. The
chamber is filled with a rarefied gas so as to generate a plasma
between the electrodes, one of which is the target and the other is

---

[6]When a deposited film has the same thickness over steps, trenches, and holes of the
substrate, we say that the film shows a good "step coverage" or "conformability."
After the substrate is already patterned, the substrate surface has various concave
and convex topographies on the surface. It is extremely important to cover the entire
surface conformably. Figure 4.8 shows different step coverage. (c) shows the best
step coverage in these three.

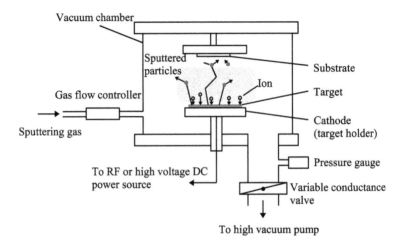

Figure 4.9 Sputter deposition apparatus.

the substrate pedestal. The target functions electrically as a cathode so that the ions in the plasma hit the target. The ion impact sputters the atoms of the target. The distance between the target and the substrate is usually between a few cm and 20 cm. The substrate is placed on the pedestal that is usually electrically grounded and functions as the anode. Figure 4.10 shows an industrial sputter tool used for LSI manufacturing. In such advanced tools, a multichamber system is used, where 2 or 3 or more deposition chamber are connected to the central transfer chamber and a robot handles a wafer. When different targets are used, different films are deposited on the same wafer. When the same material is installed to each chamber, the productivity (throughput) increases, as the number of wafers processed per unit time increases.

4.3.1.1 Sputtering phenomenon

Have you ever thrown a small pebble on gravel? If you throw normally, the stone will bounce and some pieces of the gravel will change their position. If you throw with all your might, you will flick more pieces. If you fire a pebble with something like a cannon, the stone will fly very fast, and the pebble will get buried in the gravel and will flick a lot of pieces during this process. This will be the case

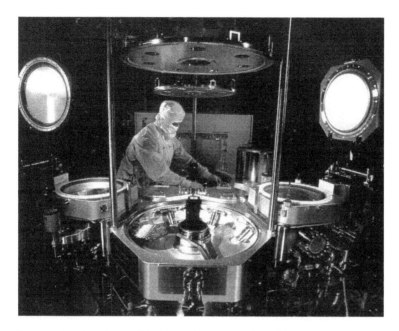

Figure 4.10   An advanced industrial sputter tool used for IC manufacturing (Endura®, Applied Materials, Inc., http://www.amat.com). Disks at both sides are targets. The hexagonal component at the center is a wafer-transfer chamber equipped with a robot.

for stacked blocks or balls. And the same thing happens when atoms or ions are thrown at a solid.

Sputtering is a physical phenomenon in which ions or neutral atoms incident on a solid surface knock out the atoms of the solid surface. When a pebble is thrown on gravel, the momentum and energy of the pebble are transferred to the gravel, resulting in the pebbles of the gravel jumping out against the gravity. As the atoms are too small, the gravity does not matter at all;[7] instead, the atoms need to break free from the binding energy between the atoms. Figure 4.11 shows the interactions between an incident particle and a solid. As an ion and a neutral particle are identical in terms of their

---

[7]The intensity of gravity is proportional to the cube of the size (length) of a particle but the atomic attraction does not depend on the size as it is very small.

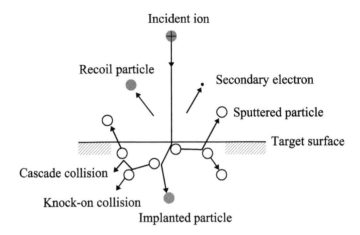

Figure 4.11   Interactions between an incident particle and a solid.

masses as well as their interactions, hereafter, we will presume that the incident species is an ion.

The incident ion kicks out solid atoms when a part of the incident energy (as well as momentum) of the ion transfers to the atoms of the target and these atoms gain energy large enough to break the bond of the surrounding atoms and become free. This means that sputtering occurs only when the energy of the incident ion exceeds a threshold energy. This process occur only on a very thin surface layer usually upto a few layers of atoms.

Below the threshold energy, the ion will just bounce back (elastic recoiling) or hit the target atoms just to vibrate them (thermal energy transfer).

When the incident energy is very high (more than 10 keV), the incident ion collides with the target atoms repeatedly—so called cascade collision—and bury deep in the surface (ion implantation).

The energy of sputtered (kicked out) particles is about 10–50 eV. This energy is 100 times larger than that of thermally evaporated particles. When the atoms having such a high energy arrived at the substrate surface, they do not "cool down" immediately, but can migrate on the surface, retaining some amount of energy. This helps in improving the film quality and step coverage.

Dominant factors that determine the film deposition rate in sputter deposition are

(i) Incident ion flux on the target
(ii) Sputtering yield
(iii) Incident flux of sputtered particles on the substrate

The incident ion flux J_{ion} depends on plasma characteristics and the ion current density is expressed by

$$J_{ion} = \sqrt{\frac{1}{m_{ion}} \frac{V^{3/2}}{\ell^2}} \qquad (4.20)$$

where V is the voltage drop at the sheath and ℓ is the sheath width. The voltage drop V depends on the bias voltage and plasma power and the sheath width depends on the pressure, and therefore, the deposition rate varies when plasma operation conditions are changed, as already discussed in Section 3.5. Next, we will discuss sputtering yield and the incident flux of sputtered particles on the substrate.

4.3.1.2 Sputtering yield

Sputtering yield η is the average number of atoms sputter-released from the target by one incident ion.

$$\eta = \frac{\text{Number of sputtered particles}}{\text{Number of incident ions}} \qquad (4.21)$$

The sputtering yield depends on the (1) ease of collision between the incident ion and the target atoms (more specifically, collision cross section, atom density, and crystal structure); (2) strength of chemical bonds of the target atoms; (3) degree of energy transfer from the incident ion to the target atoms.

An atom consists of a nucleus surrounded by electrons that are bound to the nucleus. As the nucleus is extremely small compared to the atom, an ion traveling with a moderate motion energy (100–1000 eV) observes a region as large as the electron orbital only as the collision target. The orbitals of valence electrons are widely expanded due to weak binding with the nuclei and, therefore, the valence electrons do not possess enough power (stopping power) to alter the trajectory of the oncoming ion. For these reasons, the sputtering yield is proportional to the area of the ions of the element consisting of the target, or to the square of the ion radius, r_i.

Another important parameter is the chemical bond strength. The heat (enthalpy) of vaporization is a good guide of the bond strength U. It is well known that the heat of vaporization is correlated to the melting point or boiling point T_b of the element. It is natural that the sputtering yield decreases as the bond strength between the atoms of the target increases. Therefore, the sputtering yield is supposed to be inversely proportional to T_b, i.e., $1/T_b$ can at least be used as a guide. From this discussion we can expect

$$\eta \propto \frac{r_i^2}{U} \propto \frac{r_i^2}{T_b} \tag{4.22}$$

Figure 4.12 shows the relationships between the sputtering yield and r_i^2/T_b. Despite the simplicity of our model, we can observe very good proportional relationships, proving the validity of the model.

According to the collision theory described in Section 3.3, the energy transferred from the incident ion to the target atom upon a collision is found to be proportional to

$$\frac{M_i M_t}{(M_i + M_t)^2} E \tag{4.23}$$

where M_i is the mass number of the ion and M_t is the mass number of the target element.

Finally, we obtain the function of the sputtering yield.

$$\eta(E) \propto \frac{r^2 M_i M_t}{(M_i + M_t)^2} \frac{E}{U} \tag{4.24}$$

In Fig. 4.12, we observe that sputtering yield is dependent on the incident energy and the kind of ion, which is expressed in the above equation.

r_i and U or T_b are known to show periodicity of atomic number. Therefore, the sputtering yield shows some periodicity of atomic number of the incident ion and target element.

The above equation needs to be slightly modified in an energy range slightly above the sputtering threshold energy or in a high energy range above 1 keV. η is maximum usually at tens of keV; and above this, ion implantation becomes a dominant process and the sputtering does not occur preferentially.

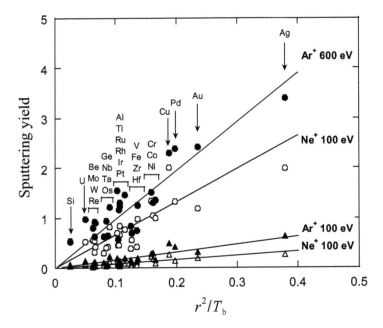

Figure 4.12 Relationships between sputtering yield and r_i^2/T_b. Symbols are grouped arbitrarily when overcrowded. The elements in a group are listed in order of their atomic numbers, and thus do not correspond to the order of the symbols.

4.3.1.3 Solid angular distribution of sputtered particles

Sputtered particles are released according to the cosine law (p. 24) so that the particles are released perpendicular to the target at the maximum probability. Materials having a higher sputtering yield (such as Au, Pt, and Cu) show good agreement with the cosine law in a solid angular distribution. On the other hand, materials with a lower sputtering yield, such as Mo and Ta, are known to show a maximum at a tilted direction, which is known as under-cosine phenomenon.

The offset from the cosine law also depends on the incident energy of the ions. A low incident energy (approximately 100–1000 eV) results in under cosine, whereas a high incident energy is known to result in over cosine. This angular distribution is often observed when a single crystal target is used.

Table 4.1 Sputtering yield at a typical incident energy for sputter deposition (0.5 keV) (The unit is atom/ion)

Gases	He	Ne	Ar	Kr	Xe	Threshold energy for Ar (eV)
Ag	0.20	1.77	3.12	3.27	3.32	15
Al	0.16	0.73	1.05	0.96	0.82	13
Au	0.07	1.08	2.40	3.06	3.01	20
Be	0.24	0.42	0.51	0.48	0.35	15
C	0.07	0.10	0.12	0.13	0.17	
Co	0.13	0.90	1.22	1.08	1.08	25
Cu	0.24	1.80	2.35	2.35	2.05	17
Fe	0.15	0.88	1.10	1.07	1.00	20
Ge	0.08	0.68	1.1	1.12	1.04	25
Mo	0.03	0.48	0.80	0.87	0.87	24
Ni	0.16	1.10	1.45	1.30	1.22	21
Pt	0.03	0.63	1.40	1.82	1.93	25
Si	0.13	0.48	0.50	0.50	0.42	
Ta	0.01	0.28	0.57	0.87	0.88	26
Ti	0.07	0.43	0.51	0.48	0.43	20
W	0.01	0.28	0.57	0.91	1.01	33

4.3.1.4 Film thickness distribution

Sputtered particles "fall" and accumulate on the substrate so as to form a thin film. When a target-substrate spacing is large, the film is deposited over a wider area than the substrate, which results in a decrease in the deposition rate. Contrarily, when the target-substrate spacing is small, the more the atoms that arrive at the substrate, the higher is the rate of the deposition, whereas the film thickness becomes more non-uniform.

The film thickness distribution of a sputter-deposited film can be obtained in the same way as we studied in vacuum evaporation in Section 4.2.1.5 (p. 101). In the case of vacuum evaporation, the evaporation source was treated as a point source, as the distance between the source and the substrate is generally very large. In contrast to this, in the case of sputter deposition, the areal spread of the target should be considered. In the calculation, the target area is considered as the aggregation of point sources, and the film thickness distribution is calculated for each source using

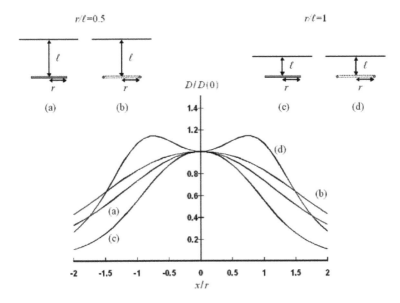

Figure 4.13 Thickness distributions for disk and circular sources.

Eq. (4.17). The superposition of the distributions gives the thickness distribution. More generally, multiple integral [Eq. (4.18)] is to be used. For a disk target shown in Fig. 4.13 (a) and (c), by setting $x_1 = r/\ell$ and $x_2 = x/\ell$, we obtain

$$\frac{D(x)}{D(0)} = \frac{1+x_1^2}{2x_1^2} \left[1 - \frac{1+x_2^2 - x_1^2}{\sqrt{(1-x_2^2+x_1^2)^2 + 4x_2^2}}\right] \qquad (4.25)$$

Magnetron sputtering in which a circular magnet (describe later) is used, a circular vapor source is used as a model. A simple model is a circle having a radius of r and zero width as shown in Fig. 4.13(b) and Fig. 4.13(d), and the solution for this model is obtained as

$$\frac{D(x)}{D(0)} = [1+x_1^2]^2 \frac{1+x_2^2 + x_1^2}{[(1-x_2^2+x_1^2)^2 + 4x_2^2]^{3/2}} \qquad (4.26)$$

The plots in Fig. 4.13 are thickness distributions calculated using the above equations. In the case of a disk target, the thickness uniformity improves as the target-substrate spacing ℓ increases, whereas the circular target shows more complicated dependences, suggesting a possibility of wider area deposition with a good film

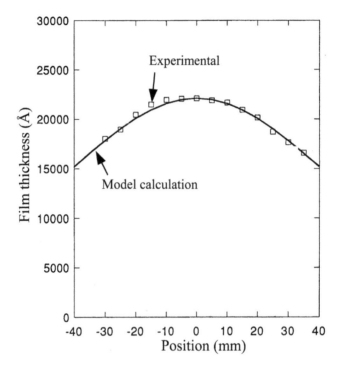

Figure 4.14 Experimental and calculated film thickness distributions.

uniformity [between (b) and (d)]. It is noted that the thickness is normalized to 1 at the film center. Thickness at the center always increases as ℓ decreases.

Figure 4.14 shows an actual film thickness distribution obtained by using a circular DC magnetron sputtering target along with calculation. We can see very good agreement which supports our discussion.

4.3.1.5 Properties of sputter-deposited films

Under usual sputter deposition conditions, the mean free path of gas molecules is of the order of cm. The target-substrate spacing is a few cm to several tens of cm so that the particles released from the target arrive at the substrate surface after colliding with gas molecules several times. The collision is a purely stochastic event;

and therefore, the majority of sputtered particles "forget" the initial direction of motion when sputtered and fall on the substrate surface from random directions, especially when the operation pressure is high. Generally, since the sputtered particle is heavier than the molecule of the sputtered gas, the energy transfer upon a collision is not very significant. Therefore, even after multiple collisions, these particles still keep the initial kinetic energy even when they arrive at the surface, at least compared to vacuum evaporation. For this reason, the arrived atoms can migrate on the surface until they are captured by lattice sites or until they grow up a nucleus of a crystal grain, or the arriving atoms can even transfer their energy to the already-deposited atoms. When such surface diffusion is significant, crystal grains grow laterally so that the formed film consists of large smooth grains. This dependence of the film structure on the deposition conditions are widely accepted as the Thornton zone model. Details of the Thornton model is described in Section 5.4.2.

A surface behaves to minimize its surface area to minimize the surface tension, such as the water droplets that become round to minimize the surface area. Active migration and diffusion of surface atoms means that the surface can change its topography so as to minimize its surface area at least locally. This improves the step coverage. Therefore, sputter deposition provides better step coverage than vacuum evaporation.

To improve the film step coverage is one of the most important subjects in film deposition engineering. Various modifications are implemented to industrial sputtering to meet this requirement. Typical approaches are

(i) High-temperature sputtering—the deposition temperature is increased as high as the melting point of the depositing material so as to induce active surface migration. Post deposition high-temperature annealing is an alternative to have the same effect.

(ii) RF bias sputtering—the substrate is negatively RF-biased so that the ions are attracted to hit the lower surface and bottom of steps. As a result, the already-deposited film is sputtered away and redeposited at the sidewall, leading to an improvement in step coverage.

(iii) To decrease the operating pressure so as to improve the directionality and linearity of the motion of atoms.

These approaches are often used together.

4.3.2 Sputter Deposition Apparatus

4.3.2.1 DC sputter apparatus and RF sputter apparatus

A DC sputter apparatus is designed based on the DC glow discharge system (p. 79, Fig. 3.11) and an RF sputtering apparatus used the RF plasma discharge system (p. 85, Fig. 3.14). A target is placed on the cathode and a substrate is placed on the anode, and a glow discharge plasma is generated between them. A blocking capacitor is used in the RF sputtering apparatus in order to apply a negative DC bias to the target (cathode). The anode is usually electrically grounded.

The kinetic energy of the ions incident on the target relax eventually to thermal energy. As the total kinetic energy is large, this thermal relaxation generates a lot of heat. The backside of the target—usually target holder or target pedestal—is water-cooled to prevent the deterioration, or melting at worst, of the target. It is noted that only electrically conductive materials such as metals can be used as a target for the DC plasma apparatus.

4.3.2.2 Magnetron sputtering

When a magnetic flux parallel to a target exists, the (secondary) electrons released from the target follow a semi-circular orbit and then return to the target, according to the magnetron motion, as already mentioned in Section 3.5.3.2 (p. 90). This magnetic configuration is easily realized by setting the magnets on the backside of the target so that the magnetic field directs from the periphery to the center of the target. This setup is called planar magnetron (Fig. 4.15).

The magnetron motion of electrons increases their motion path, which leads to an increase in the plasma density as well as the

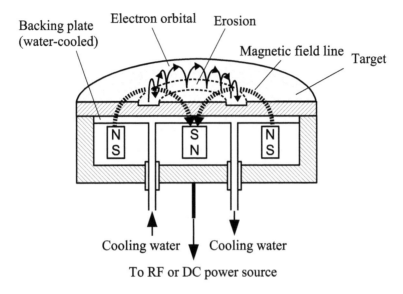

Figure 4.15 Cross-section of circular planar magnetron sputtering tool.

number density of ions. As a result, the sputter rate increases significantly and a high deposition rate of >1 µm/min can be achieved. The plasma density of this magnetron system is as high as a few percentage points and significantly large compared to that of usual plasmas without magnetic enhancement. The magnetron system decreases the plasma operation pressure due to the ionization enhancement effect (10^{-3}–10^{-1} Pa). Most of the practical sputter deposition apparatuses use the magnetron system.

One drawback is the local erosion of the target. The erosion, or local sputter etching, occurs at a region where a magnetic flux is parallel to the target surface so that the plasma density is high. The erosion forms a circular groove on the target surface; in this sense, the target use efficiency is not very high. Ferromagnetic materials, such as Fe and Ni, are not preferable for use, because the magnetic flux penetrates through the target and therefore the magnetron effect is limited. To prevent this, a thin target is used when using these kinds of materials but this reduces the lifetime of the target.

Table 4.2 Physical properties and operation conditions of plasmas

	DC, RF sputter	Magnetron sputter	Ion beam sputter	ECR, ICP sputter
Pressure (Pa)	0.1–100	0.01–10	10^{-3}–10^{-2}	0.01–0.05
Ionization	10^{-5}–10^{-4}	10^{-5}–10^{-3}	—	10^{-3}–10^{-1}
Plasma density (m^{-3})	10^{15}–10^{16}	10^{15}–10^{17}	—	10^{17}–10^{18}
Magnetic enhancement	No	Yes	No	Yes (ECR)
Note	Metal targets only for DC	either RF or DC used	—	Low pressure, high density

In advanced sputter tools, higher density and large-area plasmas such as ECR and ICP are also employed. Table 4.2 shows features and operation conditions of different plasmas.

4.3.3 Applied Sputter Deposition

4.3.3.1 Reactive sputtering

Reactive sputtering is a way to deposit compound films, such as nitrides and oxides through chemical reactions between sputtered particles and plasma gases. Atoms released from the target are mono-atoms/elementary substances and chemically reactive. In usual sputter deposition, inert gases, such as Ar are used as a plasma gas. When a reactive/non-inert gas is used—usually mixed to the main plasma gas (Ar)—the added gas decomposes or forms radicals in the plasma, and the chemical reactions with the sputtered particles are promoted so as to form a compound film.

N_2, O_2, NH_3, and CH_4 are often used as a reactive gas, which function as sources of N, O, N, and C, respectively. Nitrides such as TaN, AlN, BN, and Si_3N_4; oxides such as Al_2O_3, BeO, SiO_2, TiO_2, Ta_2O_5, HfO_2, and MgO; carbides such as TiC, SiC, and other various componds can be deposited.

For instance, when depositing TaN, a metal Ta target is used and N_2-added Ar is used as a plasma gas. In many cases, elementary metal targets are used and a reactive gas is added. This provides higher deposition rate because the sputtering yield of metals is high and better composition tunability by varying the plasma conditions, such as the partial pressure of the reactive gas, the total pressure, the plasma power, and the substrate temperature. Compound targets can also be used, especially for light element compounds such as SiO_2 and BN. As the sputtering yield of such materials is usually low and the addition of a reactive gas is still required to ensure chemical stoichiometry, such materials are not favorably deposited by sputtering unless other better techniques are available.

As understood from the above, reactive sputtering is a versatile technique to fabricate compound films. Dense and high-quality films can be deposited, compared to other deposition techniques, when the deposition conditions are tuned properly. Chemical compositions and structures are also tunable. We have to know, however, that the step coverage is poor compared to chemical vapor deposition.

4.3.3.2 Deposition of alloys and compounds

When an alloy or a compound target is used, the film of the same material is deposited, whereas its composition is not always identical. This is because the sputtering yield of every element is different. An element having a high sputtering yield is preferentially entrapped in the film. When the concentrations of the alloy elements are small, the film composition becomes close to that of the target in many cases.

The target material is sputtered into atoms and then those atoms form chemical bonds and are stabilized again. If the bond formation occurs in the gas phase, or before reaching the depositing film surface, this stable species is not caught in the film easily. For instance, when SiO_2 is used as a target, the deposited films have usually less oxygen concentration than the target. To obtain the stoichiometric composition, O_2 gas is added to the sputter gas.

In order to obtain films having different compositions, different targets that correspond to the desired compositions are needed. Obviously this is not practical especially when the number of targets are large, as targets—especially alloy and compound targets—are expensive and target exchange is burdensome.

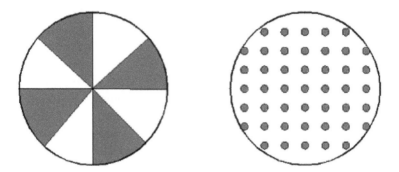

Figure 4.16   Combinatorial targets for depositing alloys and compounds.

One solution is to use a combinatorial target shown in Fig. 4.16. In this method, a target is composed of segments of different materials, or chips of alloy elements are directly placed on the main target. In order to achieve good uniformity, the substrate or substrate holder is rotated.

Another approach is to use a multiple target system, where different materials are sputtered using targets of different materials

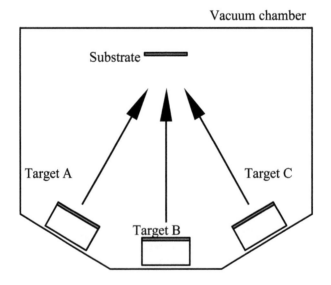

Figure 4.17   Multiple target system.

Figure 4.18   3-cathode sputter system.

at the same time, as shown in Fig. 4.17, where three different materials are sputtered using a fixed single substrate. In one variation, all the sources are operated at the same time, and the substrate is exposed to each target sequentially. This also forms an alloy film unless the rotation speed is very low. Figure 4.18 shows an actual multi-cathode sputtering system that has a shutter at each port.

4.3.3.3 Ion beam sputtering

In the sputtering systems explained so far, the target is in contact with the plasma, or the target is a part of the plasma-generation

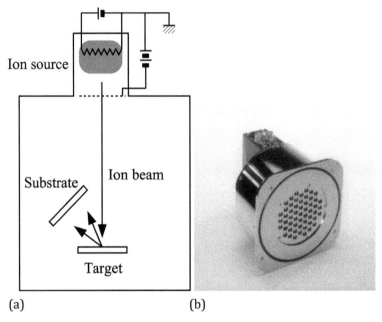

(a)   (b)

Figure 4.19 Ion beam sputter deposition system (a) and ion beam source (b), adapted from a catalogue of Commonwealth Inc.

system. If the ion source is operated independently, the target material can be sputtered by irradiating the target with accelerated ions. A typical ion energy used for this purpose is 0.5–2.5 keV. The sputtered particles deposit on the substrate located near the target (Fig. 4.19). This process is called ion beam sputter deposition or ion beam deposition, especially when mentioning the method.

The operating pressure of typical glow discharge plasmas are not very low. Therefore, the plasma gas can be entrapped in the film as an impurity. On the contrary, ion beam sources are operated at a lower pressure of 10^{-3}–10^{-2} Pa, which leads to the lowering of impurity contamination.

Another advantage is the high kinetic energy of the incident ions. In a glow discharge sputtering system, the high (10–50 eV) energy of the sputtered particles is mostly lost during successive collision with the sputtering gas before reaching the substrate surface. In ion beam deposition, due to the low operating pressure and thus a low collision probability, the original energy is maintained until the sputtered particles arrive at the surface. This promotes nucleation and growth of crystal grains and improves crystallinity.

However, due to the low operation pressure, the film deposition rate is much lower than that of glow discharge sputtering. For this reason, the ion beam source is not popularly used as a dedicated deposition aid.

Coffee Break: Grove and Edison

The research and development of electric lamps, such as arc lamps and filament lamps was carried out intensively in the middle of the 19th century. The development of a practical light bulb using a bamboo filament by a US inventor T. A. Edison (1847–1931) is known as a remarkable event (1879).

W. R. Grove (1811–1896), a British scientist, was researching on arc lamps when he invented the "safe" filament lamp in 1845. This lamp used an electrically heated Pt spiral filament, encased in a glass bulb filled with a gas. He also made remarkable achievements in the development of electrical cells. The cells and arc lamps were used for early lighting systems.

Later, Grove discovered the deposition of an electrode material on a wall after the electric discharge of a low pressure gas—that is, sputtering.

What about Edison? He carried out intensively the development and industrialization of electric sources such as rechargeable batteries and power generators. In addition to them, he developed vacuum evaporator and sputtering apparatus and filed for patents (1900). These apparatuses were used to coat wax cylinders for phonographs with gold. At present, PVD coating is extensively used for fabricating recording disks such as CDs and DVDs; is it not an interesting analogy?

Problems

(1) Describe the relationship between vaporization rate and deposition rate in vacuum evaporation.

(2) An Al film is deposited by vacuum evaporation. The source temperature is 1000 K and the pressure during the evaporation is 1×10^{-3} Pa. Calculate the following quantities.

    (a) Mean free path of Al atoms. Use $d = 1.4$ Å ($d = 0.14$ nm)

    (b) Average number of collisions of atoms. The distance between the evaporation source and the substrate is 20 cm.

    (c) Film deposition rate. The Al source is a droplet having a diameter of 6 mm. The mass density of Al is $2.7 \, \text{g/cm}^3$.

(3) Derive Eqs. (4.15) and (4.17).

(4) Ag and Au are evaporated simultaneously using identical evaporation cells. Calculate the evaporation temperatures of Ag and Au to obtain an Au–5% Ag alloy film. The unit of concentration is atomic (molar) percent.

(5) How can film uniformity in vacuum evaporation be improved? Describe the reasons and/or principles.

(6) Draw schematics of vacuum evaporation and parallel plate sputtering apparatuseş, and describe the working principles.

(7) Describe physical processes that occur when a solid surface is irradiated with an ion.

(8) Why are electrons released from the target in a magnetron sputtering system?

(9) Describe the reason(s) why the film-deposition rate is high when a magnetron sputtering system is used.

(10) Describe properties of films deposited by a sputtering method.

Chapter 5

Film Formation Process

5.1 Introduction

We have studied the behaviors of gas molecules and ions in terms of their roles as fabrication "tools." We understand that gases can behave very differently when we focus on their motion as particles of molecules and ions.

As discussed before, it is important to return to the motion of atoms and molecules in order to discuss materials and processes in a very small world. Material properties of thin films are mostly identical to those of bulk materials but can differ greatly. To understand the reasons, it is important to look into atomic scale phenomena. In this chapter, we will study the atomic behaviors in the early stage of thin film growth. Next, we will discuss the structures of thin films grown after they passing the early stage of growth.

5.2 Thin Film Growth

5.2.1 Atom Stacking and Development of a Film

A thin film is formed when atoms fall continuously on a substrate. In PVD, the incident atoms or ions directly form a film. In another

Micro- and Nanofabrication for Beginners
Eiichi Kondoh
Copyright © 2021 Jenny Stanford Publishing Pte. Ltd.
ISBN 978-981-4877-09-1 (Hardcover), 978-1-003-11993-7 (eBook)
www.jennystanford.com

popular and important film deposition technique of chemical vapor deposition (CVD), the atoms of a depositing film are formed through surface chemical reactions of adsorbed molecules incident from the gas phase.

This phenomenon is not confined to thin film deposition. When an NaCl crystal precipitates from a supersaturated salt water or when a molten cast ion solidifies, atoms in a phase other than solid arrive at the surface and settle at a lattice site.

Precipitates from salt water consist of a lot of small grains. This kind of structure is called polycrystal. Cast metals are also polycrystalline. Likewise, thin films are polycrystalline unless special peculiar means are employed (discussed afterward). A crystal has a structure in which a single atom or a unit of atoms is periodically repeated. A crystal lattice or simply a lattice is a set of atoms sectioned to a simple solid frame from a crystal. As atoms falling to the growing surface do not aim for an atomic position of the lattice, which is called a lattice site, rearrangement of the atoms must take place to form a crystal. Such rearrangement occurs at surfaces more preferably in the interior of the crystal. This means that the atoms move at the surface of the growing film.

When water vapor comes in contact of a cold window glass, droplets form on the glass. In our experience, the shapes of the droplets depend on the condition of the glass surface. If the glass is polished with liquid wax, the glass repels the water. The droplets cannot spread out along the surface; instead the droplets retain a spherical shape. As the wax erodes and the glass surface stains, the water spreads out and forms a film. This is because "wetting" (p. 143) between the glass and the water prevails against the surface tension of water.

The same thing occurs during the deposition of a solid thin film. The surface atoms are mobile, although not as much as liquid, the initial growth structure is determined by the surface tension and the wettability to the substrate, which either turns into droplets or becomes a continuous film (Fig. 5.1).

Let us see the initial film growth shown in Fig. 5.2. First, the atoms arrive at a bare substrate and get adsorbed. The adsorbed atoms diffuse on the surface (**surface diffusion**). These atoms stabilize when they meet other atoms and try to gain the structure

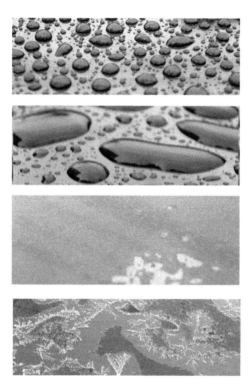

Figure 5.1 Different status of H_2O on a surface (droplets or continuous film).

of a crystal. This is called nucleation. Only two atoms are depicted in Fig. 5.2, but, in fact, a nucleus consists of a lot of atoms.

Nuclei grow by capturing the incident atoms and the surface-migrating atoms. The grown nuclei are not connected together and are discontinuous; therefore they are called islands. Furthermore, tiny islands can migrate on the surface and coalesce with each other to form a larger island. Each island grown in this manner is usually a single crystal. The islands further grow to crystal grains and the grains connect with each other to form a poly crystalline film. After a continuous film is formed, the grains of the film grow perpendicularly to the substrate. As a result, a columnar structure develops. The coalescence of the grains depends on the metallurgical conditions.

1. Condensation and nucleation

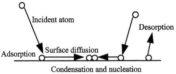

3. Coalescence and agglomeration

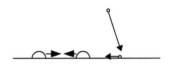

2. Nucleation and island growth

4. Polycrystalline film

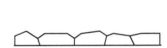

5. Grain growth

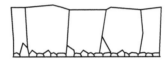

Figure 5.2 Early stage of thin film development.

5.2.2 Film Surfaces

On a growing surface, atoms arriving at the surface form an aligned structure. Let us assume an ideal case where a perfect crystal is growing. In the case when the crystal surface is perfectly flat, the film can thicken only by stacking a sheet of atoms repeatedly; obviously, this does not happen. In the second best ideal case, we assume that one layer of atoms is growing laterally on the already-grown perfect surface. This is depicted in Fig. 5.3. The edge of this atom layer is called a step, and the top surface of the layer is called a terrace. If a step is perfect, it will be straight. Next we assume that the edge has a kink, where the edge is slid by one line of atoms. The surface-diffusing atoms finally get captured at a kink and the terrace edge elongates by one atom. The first kink is formed when an atom is adhered to the flat terrace. The crystal grows by repeating this process, and this is called Kossel's model.

The number of **nearest neighbors** of a metal atom at a close-packed surface is 3–9, as shown in Fig. 5.4. The number of nearest neighbors of a lattice atom, or an atom inside a bulk crystal, is 12,

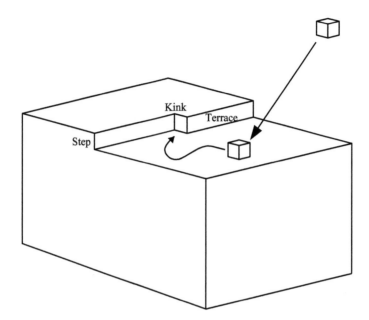

Figure 5.3   Crystal growth on an ideal surface (Kossel's model).

and that of an adsorbed atom (adatom) is 3. Therefore, the energy
needed to remove an adatom from the surface is to be 1/4 of that
of a lattice atom. The solid phase is the most stable phase when a
crystal grows, and therefore, having 12 nearest neighbors is the most
stable. This means that atoms having a smaller number of nearest
neighbors are more unstable and they have excess energy.

In a solid, diffusion atoms are formed when an atom being
vibrated by thermal energy jumps from its stable site to the next
stable site. To do this, the atom should overcome an energy barrier,
and the amount of energy needed to cross this barrier is called
activation energy. Lattice diffusion energies of metals are 1–2 eV.
The surface atoms possess excess energy and are more unstable,
which reduces the barrier energy. Activation energies for surface
diffusion are about 1/4 of those for lattice diffusion. What needs
to be insisted here is that atoms coming from the gas phase to the
surface are very mobile because they are not surrounded by other
atoms.

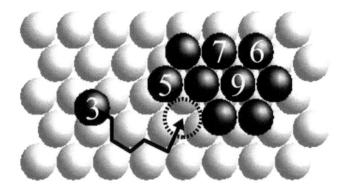

Figure 5.4 Number of nearest neighbors of atoms in a closed-packed surface. Outlined figures indicate the number of nearest neighbors.

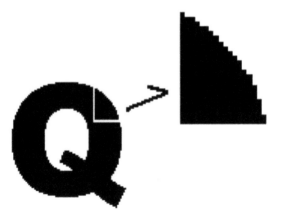

Figure 5.5 Viewing microscopically, a round surface consists of a lot of flat planes.

An atomically flat surface is an idealized model and an actual growing surface has many steps. A number of steps appear even on a low-index surface obtained by cutting or cleaving a single crystal if the cut plane is offset by a very little angle from the ideal index. This means that a round surface consists of a lot of steps (Fig. 5.5).

Figure 5.6 shows steps on a growing surface observed from an atomic force microscope (AFM). The step height is approximately a few nm.

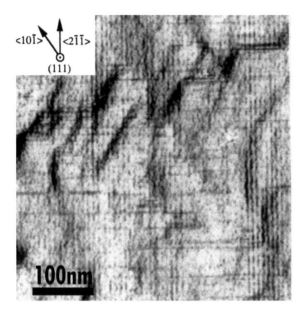

Figure 5.6 Steps on a homoepitaxially grown CVD diamond surface.

Even if a number of steps are present on a surface, the growth is terminated when all the steps sweep out the surface perfectly. To continue the growth normal to the surface, new steps should be generated continuously. This occurs when a new single or multilayer atom stack is generated on the surface. The former is referred as a two-dimensional nucleus and the latter is referred as a three-dimensional nucleus. Nucleation is discussed in detail in the next section. Screw dislocations terminating at the surface leave a step. This step advances helically while centering on the step itself, which enables continuous film thickening, similar to ascending spiral stairs (so-called BCF theory).

5.3 Nucleation

As discussed so far, a crystal grows through stacking of atoms that are captured at the surface. Then, how is the first atom layer produced?

5.3.1 Equilibrium Theory

5.3.1.1 Homogeneous nucleation

Here we discuss the case that a gas transforms to a solid without a substrate. The gas pressure is assumed to be constant.

The substance that is formed first is a very small "seed" of a crystal—a nucleus. The nucleation theory discusses whether a cluster of gathered atoms becomes stable as a nucleus or resolves to the gas phase.

When a gas transfers to a solid, the excess energy is released. This means that the total energy of atoms of the solid is greater than that of the gas by the amount of chemical bonding energy. (A solid has more chemical bonds.) The surface is part of the solid but it has excess energy as described earlier.

The problem of nucleation is a balance between the reward of energy for forming a volume and the penalty of energy for forming a surface. The nucleation theory based on the changes in **free energies** is called capillary theory.

We denote the change in energies for surface formation by ΔG_s (J/m^2) and for solid formation, or deposition, by ΔG_v (J/m^3). ΔG_s is always positive, meaning that the surface formation is not preferred. ΔG_v is negative (under the crystal growth condition). That is, we are assuming the condition where only the solid, or three dimensional linkage of atoms, is thermodynamically stable. The total change of the free energy is

$$\Delta G = V \Delta G_v + A \Delta G_s \qquad (5.1)$$

where V and S are the volume and surface area of the nucleus, respectively. For a spherical nucleus having a radius of r, Eq. (5.1) is

$$\Delta G(r) = \frac{4}{3}\pi r^3 \Delta G_v + 4\pi r^2 \Delta G_s \qquad (5.2)$$

This function has its maxima at

$$r^* = -2\frac{\Delta G_s}{\Delta G_v} \qquad (5.3)$$

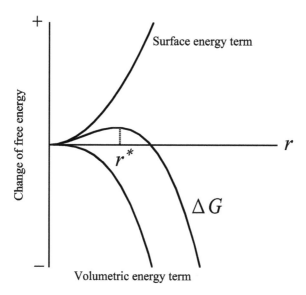

Figure 5.7 Free energy change for the formation of a spherical nucleus.

as understood from Fig. 5.7, where r^* is called **critical radius** for nucleation. Nuclei are stable when $r > r^*$ but unstable when $r < r^*$. The formation of nuclei without a surface/substrate is called **homogeneous nucleation**.

5.3.1.2 Heterogeneous nucleation

In reality, homogeneous nucleation hardly occurs, and instead, nucleation occurs on a foreign object.[1] When a thin film grows, nuclei grow on a substrate. The interfacial energy between a nucleus and the substrate, γ_i, must be taken into account. In Fig. 5.8, we assume a segment of a sphere, the radius and the contact angle of which are r and θ, respectively. Setting the areas of the base and side to S and A, respectively, and the volume to V, the change of free energy of nucleation is

$$\Delta G = V \Delta G_v + A\gamma_s + S(\gamma_i - \gamma_{sub}) \tag{5.4}$$

---

[1]Or otherwise, a foreign object works as a nuclei. Aerosol particles—dusts and specks—act as nuclei of snowflakes or ice particles (that may become rain droplets near the ground surface).

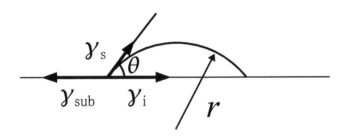

Figure 5.8 Heterogeneous nucleation.

where γ_s and γ_i denote the **surface energy**[2] and the **interfacial energy** defined per unit area. γ_{sub} is the surface energy of the substrate. $\gamma_i - \gamma_{sub}$ shows the change in free energy for interface formation.

At the perimeter of the segment, the surface/interfacial forces are balanced and are expressed by **Young–Laplace equation**.

$$\gamma_s \cos\theta + \gamma_i = \gamma_{sub} \tag{5.5}$$

By using Young–Laplace equation, calculations are carried out in a similar manner that are used to derive Eq. (5.2), and we find that the critical radius r^* is the same as in the case of the homogeneous nucleation of Eq. (5.3).[3] That is, the presence of a substrate critical radius does not influence the critical radius. The difference lies in the total free energy. The free energy

$$\Delta G^* = \Delta G_{r=r^*} = \frac{4\pi\gamma_s^3}{3\Delta G_v^2}(1 - \cos\theta)^2(2 + \cos\theta) \tag{5.6}$$

is much smaller than that of a spherical nuclei ($\Delta G^* = 4\pi\gamma_s^3/3\Delta G_v^2$), which means that the segment nucleus is formed with a smaller number of atoms. The substrate benefits from the energy for nucleation, which tremendously increases the possibility of heterogeneous nucleation.

---

[2] The value is identical to that of **surface tension**.
[3] Young–Laplace equation shows the condition in which the change of surface and interfacial energies are minimized. As a sphere is the shape that has the minimum surface area for a given volume, the critical radii are the same for heterogeneous nucleation and homogeneous nucleation.

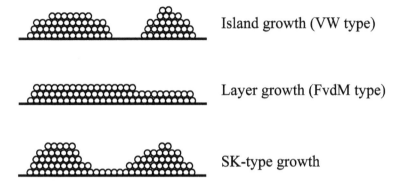

Island growth (VW type)

Layer growth (FvdM type)

SK-type growth

Figure 5.9 Film growth modes.

5.3.1.3 Surface and interfacial energy and growth mode

The Young–Laplace equation describes the so-called "**wettability**," which shows the spreadability of an adhered substance. The contact angle becomes θ small for a combination of wettable substances, and vice versa. This is highly related to the thin film growth modes.

When $\cos\theta < 1$ or when Young–Laplace equation holds and

$$\gamma_{sub} < \gamma_i + \gamma_s \tag{5.7}$$

a sphere segment is formed, and thus an early stage film growth shows an island growth mode like shown in Fig. 5.9. This is called Volmer–Weber-type growth.[4] Most of the early stage of film growth proceeds in this manner, such as in metal growth on inorganic substrates.

$$\gamma_{sub} \geq \gamma_i + \gamma_s \tag{5.8}$$

When $\cos\theta \geq 1$, since the surface energy of the substrate is high, the depositing substance prefers to form an interface only with the substrate, so that the bare substrate does not appear. The wetting substance expands and wets perfectly, leading to a layer-by-layer growth, which means that the film grows as if stacking sheets of atoms one-by-one. This mode is frequently called Frank–van der

---

[4]Equation (5.7) can be expressed as

$$\gamma_{sub} - \gamma_s < \gamma_i$$

This shows that the coexistence of the substrate and film substances is a stable state.

Merwe (FvdM) type growth. When this occurs, the strain between the depositing layer and the substrate is very small, which is an ideal condition for epitaxial growth.

Stanski–Krrastanov (SK) type growth is a special and rather uncommon mode of film growth that involves both VM and FvdM modes. The surface is covered with a few monolayers of atoms and then the island growth starts. The initial layer-by-layer growth occurs due to the strong attractive interaction between the substrate and the deposition substance. The imperfectness in lattice matching induces a large strain in the depositing layer. This increases the surface energy of the film which arises due to the nucleation on the surface of the depositing layer so as to decrease the high-energy surface. The SK growth is known to occur on the growth of some metals on semiconductor.

5.3.2 Kinetics of Nucleation

The capillary theory is an equilibrium theory, which means that time is not taken into account. Kinetics or kinetic theory is the theory to treat how much a process can proceed in a given time. As kinetic theories are complicated and elaborate, only a brief introduction is given here.

5.3.2.1 Adsorption and desorption

Incidence and attachment of gas molecules to the solid surface is called **adsorption** and detachment from the gas environment is called **desorption**.

The atoms that arrive at the surface do not directly occupy lattice sites but they migrate or diffuse on the surface for a while to find a proper site just by chance (Fig. 5.3). At the onset of nucleation, no lattice sites are present, and the migrating atoms also meet by chance. They then form a "molecule," and this molecule catches more atoms and forms a cluster. Even if the size of the cluster exceeds the critical radius, it can exist as a stable nucleus.

Desorption is a thermally activated process that breaks the weak bond between the surface atom and the surface. Setting the vibration frequency of the adsorbed atom to ν_s and the desorption energy to

$-E_{des}$, the frequency of desorption becomes $\nu_s \exp(-E_{des}/T)$. And therefore, the residence time of the adsorbed atoms on the surface will be

$$\tau_s = \frac{1}{\nu_s} \exp(E_{des}/T) \qquad (5.9)$$

Furthermore, when more than two atoms form "molecules," they become more stable than single atoms. For this reason, clusters made of a lot of atoms can survive.

5.3.2.2 Rate of nucleation

The clusters grown through this process become critical nuclei. For the sake of simplicity, we employ the equilibrium theory of formation of critical nuclei and then discuss kinetic processes of the growth of nuclei.

When one more atom is added to a critical nucleus, the nucleus becomes stable and grows continuously. Therefore, the growth rate is proportional to the number of atoms incident on the surface. The number of atoms arriving at the nucleus (atoms/s) is flux of surface-diffusing atoms × lateral area of nucleus, and thus the nucleation density $\dot{I}$ (1/s) is

$\dot{I}$ = critical nucleation density × flux of surface-diffusing atoms

   × lateral area of nucleus $\qquad (5.10)$

Next, we need to discuss each term.

First, we obtain the number of critical nuclei per unit area under equilibrium condition $(1/m^2)$. The rate of nucleation should be proportional to the number of sites N_s where nucleation can occur on the surface. Similarly, we define the number of sites N^*, where nuclei annihilate, and the annihilation rate is obviously proportional to N^*. Under equilibrium condition, the nucleation and annihilation rates are the same, and from this, we know that

$$N^* = N_s \exp\left(-\frac{\Delta G^*}{kT}\right) \qquad (5.11)$$

where ΔG^* is the change of free energy for nucleation.

Next, let us discuss the flux of surface-diffusing atoms, J. The growth species (atoms or molecules) incident from the gaseous environment stay on the surface for a period of τ_s. By setting the partial pressure of the growth species in the gaseous environment to p, the incident flux per unit area becomes $p/\sqrt{2\pi mkT}$ ($1/m^2/s$). The species will desorb in τ_s, and the number of atoms residing on the surface ($1/m^2$) is

$$\text{adsorbate surface density} = \tau_s \frac{p}{\sqrt{2\pi mkT}} \qquad (5.12)$$

per unit area.[5]

Diffusion is a thermally active process, and the frequency of the surface-diffusion motion of atoms per unit time is

$$\nu_s \exp\left(\frac{-E_{\text{diff}}}{kT}\right) \ (1/s) \qquad (5.13)$$

where E_{diff} is the activation energy for surface diffusion. By multiplying Eq. (5.12), the number of atoms that surface-diffuse per unit time is

$$\tau_s \frac{p}{\sqrt{2\pi mkT}} \nu_s \exp\left(\frac{-E_{\text{diff}}}{kT}\right)$$

and, therefore, we obtain the flux of surface-diffusion atoms J as

$$J = \frac{p}{\sqrt{2\pi mkT}} \nu_s \exp\left(\frac{E_{\text{des}} - E_{\text{diff}}}{kT}\right) \qquad (5.14)$$

Finally, the lateral area of a nucleus A is $A = 2\pi r^* a \sin\theta$ where a is an equivalent height of a single atom layer. Therefore, we reached the equation of $\dot{I}$

$$\begin{aligned}
\dot{I} &= N^* J A \\
&= \frac{2\pi r^* a \sin\theta \, p N_s}{\sqrt{2\pi mkT}} \exp\left(\frac{E_{\text{des}} - E_{\text{diff}} - \Delta G^*}{kT}\right)
\end{aligned}$$

$$(5.15)$$

The most sensitive term in Eq. (5.15) is obviously the exponential term. The change of free energy for condensation $\rightleftharpoons$ vaporization of vapor, δG_v, is

$$\Delta G_v = -\frac{kT}{\Omega} \ln \frac{p_v}{p_s} \qquad (5.16)$$

---

[5] Adsorption and desorption of atoms occur continuously, and only a small numbers of those atoms are used in the growth of nuclei. Therefore, we can safely assume (number of adsorbing atoms) = (number of desorbing atoms).

where p_v and p_s are the partial pressures of the growing species (vapor atoms in this case) in the gas phase and at the surface, respectively, and Ω is the volume of an atom. When p_v is much larger than p_s, the vapor is in oversupply against the equilibrium,[6] which increases ΔG and thus increases the nucleation rate significantly. When the nucleation density is large and the surface diffusion lengths are larger than the nuclei spacings, most of the adsorbed species are captured by the already existing nuclei; and therefore, further nucleation does not take place, leading to the formation of a continuous film.

Atoms meet together and form a cluster, and then nuclei develop. In atomic theory, behaviors of each atom are to be discussed. Statistical and kinetic models that deal with the attachment/detachment of surface-diffusing atoms to/from a cluster or "molecule" can be established. Due to the brilliant innovation of computers and numerical methods, molecular dynamic and molecular orbital calculation are now being popularly employed. They are much beyond of the scope of this book, but it is important to emphasize that equilibrium theory is still widely used as a powerful tool when discussing thin film growth.

5.4 Development of Film Microstructures

5.4.1 Island Growth and Coalescence

It is a common observation that liquid droplets coalesce and form a larger droplet. Islands grown from critical nuclei coalesce in a similar manner. Island coalescence was observed directly from *in-situ* deposition experiments carried out in an electron microscope. (Fig. 5.10).

First, we will study how surface morphology influences the surface energy. Figure 5.11 is a two-dimensional lattice that depicts the relationship between the surface topography and the number of nearest neighbors. Seeing this lattice from a direction perpendicular to the plane of the paper, the number of nearest neighbors of the

---

[6]i.e., supersaturation $(p_v - p_s)/p_s$ is large.

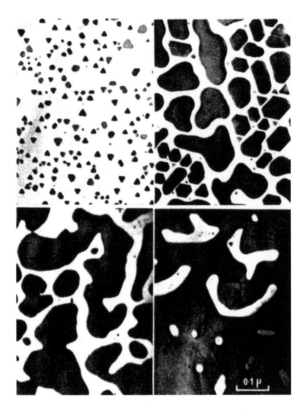

Figure 5.10 Early stages of thin-film growth observed by using transmission microscopy (Pashley).

flat surface is 5. The number of nearest neighbors are smaller on a convex plane and surfaces with a smaller curvature radius.[7] As already stated in Section 5.2.2, the excess energy is distributed to the atoms that have less number of nearest neighbors. Those atoms are more unstable, which means that atoms at a surface that has a smaller curvature radius are more unstable.

Figure 5.12 shows how islands develop to a continuous film. Smaller islands have a smaller radius of its circumference ($r > 0$), and therefore, unstable surface atoms can easily detach from the

---

[7]The radius of a circular arc (infinitesimal) that approximates a curve. The sign of the curvature radius of a convex is positive and vice versa. The curvature radius of a straight line is ∞.

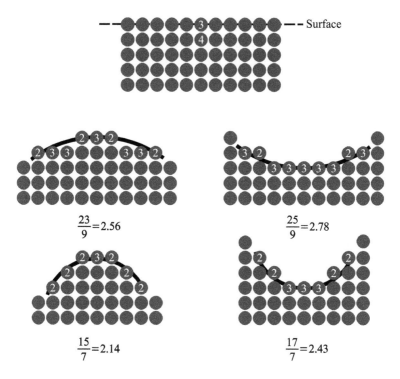

Figure 5.11 Number of nearest neighbors of the surface atoms.

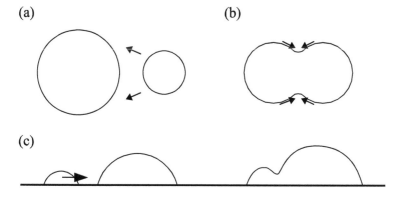

Figure 5.12 Coalescence and growth of islands.

island and migrate on the surface. Contrarily, islands that have a larger radius are more stable. Overall, atoms diffuse from smaller islands to larger islands, and as a result, smaller islands annihilate and the larger islands grow more. This phenomenon is called **Ostwald ripening**.

When islands grow, they come in contact with each other. This results in **sintering** of the islands. The contact areas of islands are constricted (concave) and have a negative curvature radius. This part has a larger surface energy than the flat surface but the addition of atoms to this part comforts the high surface energy. As surface that has a positive curvature can be stabilized by reducing the curvature, the atoms diffuse from the convex part to the concave part. This principle is identical to the coalescence of liquid droplets, in which the total surface energy (identical to the total surface tension) of the islands is minimized.

Clusters are mobile and can migrate but very slowly compared to atoms. The smaller the size and the higher the temperature, the more mobile the clusters will be. Clusters having a size up to 50–100 Å are said to be mobile. When clusters come in contact with each other, larger islands are formed through Ostwald ripening and sintering.

5.4.2 Development of Polycrystalline Film Structures

As nuclei grow to islands and to a continuous film, the thickening film shows a polycrystalline structure. The dependencies of classified structures on the deposition parameters can be depicted like a map that consist of different "zones."

The most known and widely accepted zone model is the one classifies film structures of various magnetron-sputtered metal films (a few tens–a few hundreds μm) into four zones, namely zone I to zone Ⅲ and zone T (Fig. 5.13). In Fig. 5.13, T/T_M is the absolute deposition temperature normalized by the melting point (homologous temperature).[8]

---

[8]Activation energies of thermally activated processes of different substances, such as diffusion, are generally proportional to their melting points. Therefore, thermally activated processes are said to be identical at the same homologous temperature.

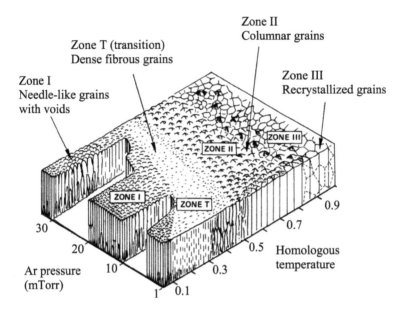

Figure 5.13 Thornton's zone model (based on Thornton's original paper), 1 mTorr = 133 Pa.

Zone II Atoms easily diffuses on a surface, when the deposition temperature is high ($T/T_M \gtrsim 0.5$). Grain boundaries are the interface between crystal grains and are energetically unstable. Atoms diffuse to minimize the total interfacial energy, and thus, the total surface area. Consequently, the grain boundaries move and the grain size increases (grain growth). After the grain growth is completed, the film thickens with keeping its grain sizes, the film exhibits a columnar structure.

Zone III At much higher temperatures, the minimization of grain boundaries occur three-dimensionally. This recrystallization result in the development of equiaxed grains.

Zone I When the deposition temperature is low, the mobility of atoms is low. The nucleation density increases whereas the grain growth hardly occurs. This results in a very fine fibrous structure. Development of a fibrous structure frequently accompanies amorphous phases.

Zone T T stands for the transition zone, where surface diffusion starts to occur significantly. Part of grain boundaries become mobile so that grain growth occurs in part, leading to the coexistence of larger and smaller grains (bimodal grain size distribution).

On the other hand, when the pressure is low, the vapor atoms travel longer without colliding with other atoms. Less atom scattering results in the formation of voids, because the protruded or convex parts formed occasionally on the surface shadow the incident atoms, known as the shadowing effect. However, the movement of atoms that reside or get adsorbed on a shadowed surface are not hindered by the incident atoms and therefore are still mobile. Therefore, on comparing at the same temperature, it is observed that the crystal growth proceeds slightly preferentially when the pressure is lower.

The zone model can be applicable to either sputtered films or evaporated films, in general. In the case of sputtering, the incident atoms have kinetic energy as high as at least a few eV when the operating pressure is low. This energy is much higher than thermal energy for evaporation (1/40 to 1/10 eV) so that the incident atoms can activate the surface atoms. This effect is similar to the temperature increase in vacuum evaporation.

As seen above, the film microstructure can be treated co-ordinately with respect to atom energy (thermal and kinetic). Figure 5.14 shows transmission electron micrographs of Al films

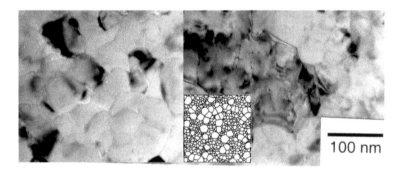

Figure 5.14 Al films deposited by sputtering at room temperature. The structures exhibit unimodal and bimodal structures depending on the deposition conditions.

deposited by sputtering at room temperature. The structures exhibit unimodal and bimodal structures depending on the deposition conditions, which clearly indicates that the grain size is dependent on the growth conditions even if the same deposition apparatus is used.

5.4.3 Epitaxial Growth

By suspending a seed crystal in supersaturated saltwater solution and growing the crystal slowly and carefully, we can obtain a large single NaCl crystal. Similarly, in thin film growth, using a single crystal substrate as a seed, a single crystal film can be grown via careful growth operation. Epitaxy is the phenomenon in which a (single) crystalline film is grown when the film lattice is matched with the lattice of the underlying layer.[9]

A lattice fit or matching means that the structure and **lattice constant** of the film are inherited from those of the substrate; this happens when the film and the substrate have very similar in-plane crystal structures and lattice constants even between different substances. A lattice fit does not always occur even between the film and the substrate of the same substance if their plane indices and/or crystal structures are different.

Epitaxial growth is the film-growth phenomenon or method in which the film and the substrate have an epitaxial relationship. To grow a layer of the same material as the underlayer is called **homoepitaxial** growth and, likewise, to grow a different substance is called **heteroepitaxial growth**. In many cases, to grow a single-crystal film all over the substrate is called homoepitaxial growth, whereas this word is frequently used when a polycrystalline film grows while inheriting the lattice structure of the substrate.

The growth of NaCl crystal mentioned above is the homoepitaxial growth. Such growth occurs in any single crystal growth, even on the growing plane of a single crystal grain of a polycrystal. For this reason, the term "homoepitaxial growth" should be understood in its technical form. In view of film-growth technology, homoepitaxial growth refers to growing a nearly perfect and defect-free single

---

[9]Combination of Greek words for ἐπι (upon, at) and ταξια (arrangement, order).

crystal on a carefully prepared substrate single crystal. Controlling the crystal plan and decreasing the number of defects are the main technological concerns. The substrate crystal is usually synthesized by the solidification growth technique.

The heteroepitaxial growth tends to occur when the lattice structures of the film and the substrate are the same and occur favorably when their lattice constants are close together. The degree of misfit, or the **misfit parameter**, is defined by

$$\eta = \frac{|a_1 - a_0|}{a_1} \tag{5.17}$$

where a_0 are a_1 are the lattice parameters of a unit cell of the film and the substrate. Figure 5.15(a) shows two-dimensional square lattices for a misfit parameter of 2%, where open symbols indicate the atoms of the substrate and solid symbols indicate the film atoms.

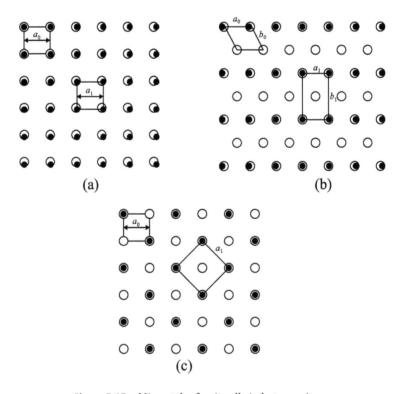

Figure 5.15 Mismatch of unit cells in heteroepitaxy.

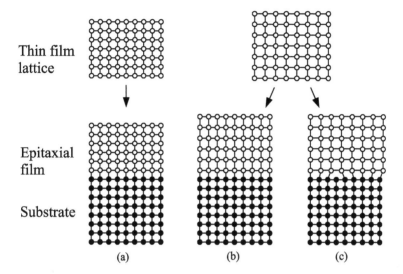

Thin film lattice

Epitaxial film

Substrate

(a)   (b)   (c)

Figure 5.16   Strains and dislocations introduced due to interfacial mismatching.

In Fig. 5.15(b), the unit cell of the substrate is rhombic (monoclinic) whereas that of the film is oblong (tetragonal or orthorhombic).[10] The b-axis of the substrate lattice fits the substrate whereas the a-axis shows poorer fitting. In another case, good fitting is obtained by lattice rotation as shown in Fig. 5.15(c).

Besides the case of perfect matching shown in Fig. 5.16(a), the epitaxy between unmatched lattices requires a way to relax the differences. Figure 5.16(b) shows an example of relaxation where the film lattice is strained to fit the substrate lattice. In the case of (c), the lattice strain does not fully relax the misfit and therefore the crystal planes of the substrate side are drawn out in appearance. This corresponds to the introduction of dislocations, and in reality, the dislocations are usually introduced in the film side. When the misfit is large, epitaxial growth does not take place at all or significant amount of interfacial defects are introduced.

The occurrence of epitaxial growth depends not only on the misfit parameter but also on the combination of the film and

---

[10] Terminology of crystal system is sometimes confusing due to three-dimensional arrangements and also partly due to historical background.

substrate materials. Qualitatively speaking, (1) the substrate surface must be very clean, (2) the surface and interfacial energies must be better balanced to promote layer-by-layer growth as shown in Eq. (5.8), (3) the growth temperature must be high enough to promote surface diffusion and island coalescence, (4) the atom/molecule flux must not be high compared to growth rate, or the supersaturation must be minimized in order to avoid unnecessary nucleation especially three-dimensional nucleation,[11] (5) The deposition atmosphere must be clean enough so as not to contaminate the growing surface.

Coffee Break: Neither Liquid or Gas or Solid

We have learnt that substances are present in either of the three states of matter—gas, liquid, or solid. However, substances do not always take such states.

A gel is a state between a solid and a liquid. A lot of gooey matters such as gelatin, jellies, and chewing gums are gels. Gels become liquid when heated without deterioration and become solid when cooled. A toy slime is a gel.

A liquid becomes a gas when heated and a gas becomes a liquid when pressurized. When the pressure and temperature exceed a critical point, a substance becomes a supercritical fluid. Supercritical fluids can act as a solvent and have high chemical reactivity. They can also be used to extract caffeine and aroma oils, manufacture nanostructures, and render toxic chemicals harmless.

By the way, is a toothpaste liquid? What about softcreams?

Epitaxial single crystal films possess ideal and inherent properties of the substance due to low defects and controlled crystal orientation. Homoepitaxially grown Si films are used to improve the performance of transistors (used to grow a layer that has a low-impurity concentration on a substrate that has a high-impurity concentration). Epitaxial growth allows to form a sharp and physically ideal interface between different substances so as to

---

[11] Supersaturation decreases as temperature increases. A high density of two-dimensional nucleation can lead to a high-quality continuous film.

control the electronic status three-dimensionally, which is used to improve the performance of electronic and magnetic devices. Light-emitting diodes and semiconductor lasers were realized only with hetroepitaxial growth technology of compound semiconductors.

Problems

(1) Set an excess energy of an isolated and free atom to 1. Obtain excess energies lost when a surface ad-atom (single atom on a surface) is captured (a) by a step and (b) by a kink as shown in Fig. 5.4.

(2) Activation energies of lattice diffusion in metals are approximately 1–2 eV. The activation energies for surface diffusion decreases to $1/4$. Why?

(3) Obtain a critical radius r^* for homogeneous nucleation.

(4) Equation (5.4) shows the free energy change for the nucleation of a spherical cap. Confirm this equation.

(5) Confirm the Young–Laplace equation in terms of the balance between surface and interfacial energies.

(6) Calculate mismatch parameters for Au//Ag, Cu//Ag, and AlAs//GaAs. // shows an epitaxial relationship and the substance in the right-hand side is the substrate.

(7) Explain the following terms:
Volmer–Weber-type growth, Frank–van der Merwe–type growth, Stanski–Krrastanov-type growth, Ostwald ripening, mismatch parameter, homoepitaxial growth, and heteroepitaxial growth

(8) Explain the origin of each zone in the Thornton's zone model.

Chapter 6

Etching

6.1 Introduction

Unnecessary part of a thin film or substrate must be removed for fabricating desired pattern and geometry. Etching is a process in which unnecessary parts of thin films are removed using chemicals and plasmas.[1] In this chapter, we focus on etching processes using gases and plasmas. Etching based on chemicals is also described because it is widely used. The principle of etching is to break the chemical bonds of an object workpiece so as to expel them to a gas or liquid phase. This core concept should be kept in mind throughout this chapter.

6.2 Classification of Etching

A variety of etching methods are available and are used according to intended uses. They can be classified as, for example,

---

[1] Sometimes used for batch processing by using chemical solutions in a narrow sense.

Micro- and Nanofabrication for Beginners
Eiichi Kondoh
Copyright © 2021 Jenny Stanford Publishing Pte. Ltd.
ISBN 978-981-4877-09-1 (Hardcover), 978-1-003-11993-7 (eBook)
www.jennystanford.com

$$
\left.\begin{array}{l} \text{Wet etching} \\ \text{Dry etching} \end{array}\right\} \left\{\begin{array}{l} \text{Physical} \\ \text{Chemical} \\ \text{Physical + Chemical} \end{array}\right\} \left\{\begin{array}{l} \text{Batch} \\ \text{Successive} \end{array}\right\} \tag{6.1}
$$

Classification shown in the first brackets is for methods, the second is for principles, and the third is for the size of an object. By removing uncommon combinations, we obtain the following classification.

$$
\left\{\begin{array}{l} \text{Wet etching} - \text{Chemical} \\[2ex] \text{Dry etching} \left\{\begin{array}{l} \text{Physical} \left\{\begin{array}{l} \text{Batch} \\ \text{Successive} \end{array}\right. \\[2ex] \text{Chemical} \\[2ex] \text{Physical + Chemical} \left\{\begin{array}{l} \text{Batch} \\ \text{Successive} \end{array}\right. \end{array}\right. \end{array}\right. \tag{6.2}
$$

There are other classification in terms of pattern topography

$$
\left\{\begin{array}{l} \text{Isotropic etching} \\ \text{Anisotropic etching} \end{array}\right. \tag{6.3}
$$

and in terms of the possibility of removing only a desired material among different materials

$$
\left\{\begin{array}{l} \text{Selective etching} \\ \text{Non-selective etching} \end{array}\right. \tag{6.4}
$$

Furthermore, etching methods can be classified on the basis of the means used to break chemical bonds (activation means). The concept of the above classifications are briefly described below and discussed in detail from Section 6.3.

Wet etching is an etching technique using a liquid chemical solutions. Contrarily, **dry etching** uses plasmas or reactive gaseous environment instead of solutions Fig. 6.1.

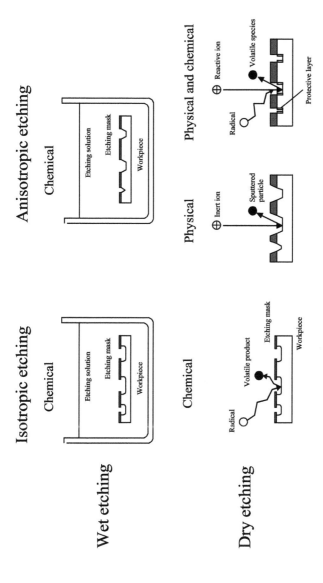

Figure 6.1 Classification of various etching techniques.

Reaction vessels are the basic requirement for wet etching because they ease scale-up and lower equipment and running costs. For this reason, wet etching is very popularly employed from laboratories, using beakers, to industries. Contrarily, dry etching requires expensive facilities such as vacuum systems and plasma generators. This limits scale-up and increases various costs, whereas dry etching allows precise fabrication and is thus widely used in microfabrication such as in IC manufacturing.

In physical etching, phenomena such as sputtering are used to remove a substance. It involves decomposing a solid to atoms and molecules, and they are expelled to a gas phase or vacuum. All physical etching techniques are dry etching.

In contrast, chemical etching uses chemical reactions to remove a substance. In the case of dry etching, volatile species are produced and a workpiece solid vaporizes to a gas phase or vacuum. In wet etching, reaction products that are soluble to aqueous solutions are formed, and a workpiece solid dissolves in the solutions.

In advanced dry etching processes, physical and chemical principles are used in combination.

In **isotropic etching** etching, reactions proceed isotropically, which means that the removal rate along the surface normal is the same over the surface. In **anisotropic etching**, the etching rate is dependent on the direction of the surface normal, which enables to etch or unetch particular surfaces. Isotropic etching is used to form sharp edges and high–aspect ratio geometries.

As shown in Fig. 6.2, in a **batch etching process**, the entire surface is exposed to the etching environment (solution or plasma), and in a **successive etching process**, etching is performed on a limited area of the workpiece and the etching position is successively moved. In batch etching, the area not to be etched is covered with an **etching mask**[2] (similar to a photoresist, see Section 6.4, p. 169), and only the uncovered area is removed. Etching chemistry and material combination are to be carefully designed to etch only the desired material without deteriorating etching mask materials. Batch etching is very commonly used in industry due to its high productivity. In successive etching, patterning is done

---

[2]Do not confuse with photomask.

Batch process   **Successive process**

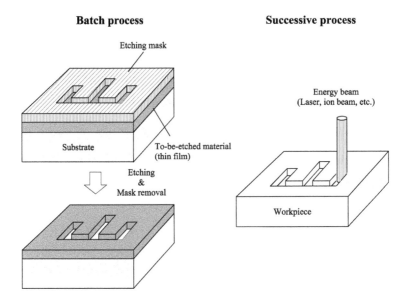

Figure 6.2 Batch etching and successive etching.

Selective etching   Non-selective etching

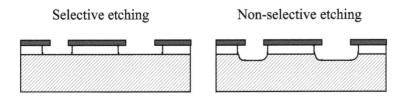

Figure 6.3 Selective etching and non-selective etching.

directly like pen drawing. The etching mask is not necessary but the removal rate is very low. For this reason, successive etching finds limited use. Correction of photomask patterns is a representative example.

When a thin film on a substrate is etched, only the film is etched without eroding the substrate, as shown in Fig. 6.3. In addition, the etching mask should not be lost during etching and its dimension must be kept as required. **Selective etching** refers to etching a particular substance exclusively. The **etching selectivity** is the ratio of the etching rates of two substances.

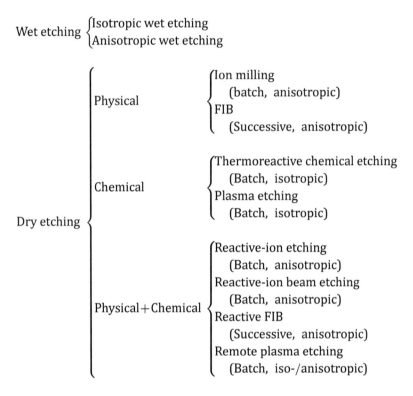

Figure 6.4 Etching techniques discussed in the book and their classifications.

Currently used etching methods will belong to any of the above categories. In this section, etching methods shown in Fig. 6.4 are described.

Requirements for etching are common, regardless of the methods employed, such as

(i) High but controllable etching rate
(ii) Good etching uniformity
(iii) Good selectivity
(iv) High precision and good isotropicity or nonisotoropicity, depending of requirements
(v) Low damage to remaining materials
(vi) Safety and low environmental load

6.3 Wet Etching

6.3.1 Isotropic Etching

Recall how a cube of sugar dissolves in still water. The cube of sugar dissolves uniformly and becomes round in shape (and finally disappears). This is because dissolution proceeds isotropically. The corners have a larger solid angle and the dissolving solute is transported to the outside environment more easily. As a result, the outside corners or convex features dissolve faster than the other portions. In actual etching, unmasked parts become round in shape inward due to the reason opposite to the case of outside corners.

Let us see the left side Fig. 6.5. As the etching proceeds uniformly, the etched thickness is the same over the entire surface. At the unmasked opening, the etched edge becomes parallel to the etching mask, whereas the edges beneath the mask will become round. In this figure, the geometry of original surface 1 changes 1 → 2 → 3 → 4 → 5 in order.

The phenomenon or sometimes degree of etching proceeding laterally under an etching mask is called side etching or etching. The etching width becomes larger than the dimension defined by the mask dimension, resulting in poor preciseness. The dimension of side etching sometimes is as large as a few μm or more. However, the side etching is not always to be ruled out, as self-supporting structures such as cantilevers and diaphragms are formed by utilizing this phenomenon. Important aqueous solutions for isotropic etching are listed in Table 6.1. Some of these need special caution when handling.

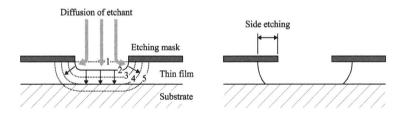

Figure 6.5   Procedure of isotropic etching.

Table 6.1 List of common isotropic wet etchants

Materials	Etchants	Compositions
Ag	$NH_4OH + H_2O_2$ (1:1)	
Al	$H_3PO_4 + HNO_3 + CH_3COOH + H_2O$	4:1:4:1
Au	$I_2 + NH_4I + H_2O + CH_3OH$	
Cr	$Ce(SO_4)_2 \cdot 2(NH_4)_2SO_4 \cdot 2H_2O + H_2O$	
Cu	$FeCl_3 + H_2O$	
Mo	$H_2SO_4 + HNO_3 + H_2O$	1:1:3
Ni	$FeCl_3 + H_2O$	
Pd	$HCl + HNO_3$	3:1 (aqua regia)
Pt	$HCl + HNO_3$	3:1 (aqua regia)
Ti	$HF + HNO_3 + H_2O$	1:1:50
W	$KH_2PO_4 + KOH + K_3Fe(CN)_4 + H_2O$	
Al_2O_3	Hot H_3PO_4	
GaAs	$H_2SO_4 + H_2O_2 + H_2O$	
Si	$HF + HNO_3 + CH_3COOH$	
SiO_2	$HF + NH_4F$	Commercially available as BHF
Si_3N_4	HF	
Glass	conc. HF	

6.3.2 Anisotropic Etching

Contrary to isotropic etching, in anisotropic etching, the etching rate depends on the etched directions. Using the example of a cube of sugar—although this does not happen in reality—the cube shape is maintained if etching proceeds anisotropically.

Is it not a wonder that dissolution proceeds in particular directions? The mechanisms of wet anisotropic etching are mainly twofold.

In the first case, the so-called crystallographic anisotropic etching, particular crystal planes remain because of their low etching rates. Different crystal planes have different atomic arrangements thus different reactivity against etching chemicals. Crystal planes are identical to "cut planes" of a crystal and exist numerously. Among them only the plane that has the lowest etching rate remains after etching, especially when the etching rate is exclusively different (**Appendix C**). Using the example of a cube of sugar again, if (100) plane remain, the cube sugar maintains its cube shape. Anisotropic etching of covalently bound crystals such as Si and GaAs are well known to occur when particular etchants are used.

In the second case, the chemical reactivity of the to-be-etched areas is controlled slightly differently from the not-to-be-etched areas by additives. Although etching itself proceeds in an isotropic manner, the difference in chemical reactivity results in the difference in etching rates and particular areas remain unetched (impurity anisotropic etching).

One practical application of crystallographic anisotropic etching is to engrave trenches and holes in the surface of a covalent crystal.

Si is a material appropriate for crystallographic anisotropic etching and is frequently used in micromachining for this reason. By using a proper etchant, the etching rate of {111} planes can be lowered greatly compared to that of {100} planes so as to remain only {111} planes. In the case of a Si wafer of which surface orientation is (100) plane, or a (100)-orientated wafer, {111} planes are inclined at an angle of 54.7° to the surface, V-shaped grooves are formed. Similarly, for (110)-oriented wafers, {111} planes are inclined at an angle of 90°, and trenches with vertical sidewalls are formed (Fig. 6.6). Table 6.2 lists some important etchant for Si isotropic etching.

Impurity isotropic etching utilizes the etching-rate dependence on impurity concentrations and is widely used in Si micromachining. The "impurity" here—often called dopant or extrinsic impurity—means elements intentionally added to Si to change the electric conductivity and to change the conduction type Si to n-type, which indicates the high free-electron concentration or p-type where the concentration of positive holes is high. Etching reactions are oxidation–reduction reactions and accompany electron or hole

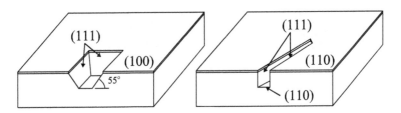

Figure 6.6 Anisotropic etching of crystalline Si.

Table 6.2   Etchants for Si anisotropic etching

Composition	Note
KOH 20–50%	20–100°C
Ethylendiamine + pyrocatechol + water (EDP)	500ml:88g:234ml
Tetramethylammonium hydroxide (TMAH) 5–20% + water	60–90°C

transfer. This is why the rate of etching depends on the conductivity types and concentrations of impurities. As this is basically an electrochemical phenomenon, an external voltage can be applied to the specimen/workpiece to control the etching rate.

The most common way of adding impurities is to use ion implantation. The use of a mask allows area-selective doping, and the doping depth can be controlled from sub-μm to a few μm by the conditions of implantation and the subsequent diffusion anneal. By doing so, only the surface not to be etched can be modified, which enables fabrication of thin beams and diagrams.

6.4 Physical Dry Etching

Sputter removal of a target material by ion irradiation was described in Section 4.3.3.3 (p. 129). By replacing the target with a to-be-etched workpiece, the surface of the workpiece can be etched. This is the principle of ion beam etching and this is a representative method of physical etching. Either batch (whole area) or local (successive) etching can be performed.

In batch etching, the whole surface of a workpiece is irradiated with ion like a shower. This is called **ion milling**. The ion sources used is basically identical to the one shown in Fig. 4.19, and the equipment is also very similar except that a workpiece is placed instead of a target.

As ion milling is a batch process, etching masks are needed to etch a pattern. An issue is that the etching selectivity of physical etching is not as high as chemical etching so that the etching mask material is concurrently removed while the unmasked part is being removed. Low–sputter yield materials are act as better etching masks, and the thickness of a mask should be carefully determined so that it is not lost while an underlying thin film is being patterned.

A photoresist is a good etching mask for physical dry etching. Photoresists are organic substances that consist of light elements and have a lower sputtering yield than that of metals. This enables larger etching selectivity against metals. Refractory metals such as W show a low sputtering yield and are frequently used as etching mask materials. When using materials other than photoresist, photolithographic processes are additionally needed as well as the subsequent chemical etching. Although the low etching selectivity is a drawback of ion milling, it is often used due to its versatility in etching various materials.

The sputtering yield is dependent on the ion incident angle θ and have a maximum at about $\theta = 40°$ ($0°$ is the normal), where the peak angle may vary with the incident energy and the material.[3] For this reason, when the ion beam is incident normal to the substrate, the etching rate is maximum in the direction tilted by this angle from the normal. As a result of this, and also due to the re-deposition of sputtered material, the sidewalls become tilted. This can be abated by tilting the ion beam angle and by rotating the substrate.

As for local (successive) etching, a focused ion beam is used. The generation of a focused ion beam is much more complicated than ion milling. After ionizing a gas, the ions are introduced to a mass-separator to select particular ions. The ions are then focused by using electrostatic lenses and the focused ion beam is applied to the surface. This method is called **focused ion beam etching** or **FIB etching** (Fig. 6.7). The FIB etching enables very fine and micro- to nanoscale fabrication, whereas the mass-separator and electrostatic focusing systems are very complicated and form a large and very expensive apparatus.

Inert gases such as Ar and Xe are used for a gas of an ion beam etching. In addition to them, vapors of metals and alloys such as Ga,

---

[3]Ions incident obliquely do not penetrate into the surface deeply; they exchange the motion momentum with atoms in this shallow region and the atoms that gain the momentum can escape easily.

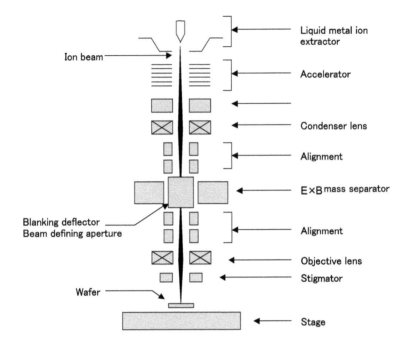

Figure 6.7   Construction of focused ion beam etching apparatus.

Si, and Au are ionized and used as an ion source.[4] For vaporization, an electric field–induced evaporation method is used where the metal vapor is ionized during evaporation. Metals that have a lower melting point (easy to handle) and higher vapor pressure (large ion current) and are chemically stable are used, and Ga (melting point 293°C) is popularly used. Alloys are used when impurity doping is needed.

As ions come straight and continuously, only the ion-irradiated part is removed; in this sense, FIB etching can be considered anisotropic etching. In a one-ion impact, 1 to 10 layers of surface atoms are removed.

---

[4] Liquid metal ion sources are used. Metal ions have a large diameter and small ion range, which allows focusing a beam onto a very narrow area of size smaller than 0.1 μm.

6.5 Chemical Dry Etching

In chemical dry etching, the etching process proceeds only chemically. To do this, gases that react with the workpiece preferentially—selectively against etching masks or other materials, if possible—must be used. Therefore,

(1) Stable reaction products must be formed.
(2) Reaction products have a high vapor pressure, which is enough to vaporize in a vacuum.

are required, which is in common for all dry etching techniques using chemical reactions. Chemical etching enables material removal at a very high rate with a proper combination of chemicals and materials.

6.5.1 Thermoreactive chemical etching

In the case of wet etching, etching reactions proceed only by dipping a workpiece in an aqueous solution heated when necessary. Similarly, under particular conditions, etching reactions can occur even by just exposing a workpiece to a reactive gas. These reactions are driven only by thermal energy, therefore, this etching process is called thermoreactive chemical etching. Here, Si etching using XeF_2, a fluorinated compound, is described.

XeF_2 is a solid at room temperature and vaporizes in a vacuum because it has a high vapor pressure. Gaseous XeF_2 reacts with Si very well.

$$Si + 2XeF_2 \rightarrow SiF_4 + 2Xe \qquad (6.5)$$

The reaction product SiF_4 is a gas therefore the reacted part vanishes. XeF_2 reacts with Si selectively. The etching selectivity of Si against SiO_2 is more than 10 thousand times and shows good selectivities against photoresists and metals, which is better than wet etching using $HF–HNO_3$ solutions. This reaction is not sensitive to crystal orientation and the process preciseness is not very high compared to other dry etching techniques. For these reasons, this technique is frequently used for sacrifice layer etching.

Wet etchants often do not penetrate into small structures due to surface tension and viscosity. Stirring and diffusion affects macroscopic and microscopic uniformity of chemicals and byproducts. Gas bubbles formed by reactions behave like an etching mask, which impede the supply of etching chemicals. Thermoreactive dry etching is free from these issues and provides better etching accuracy.

6.5.2 Plasma Etching

XeF_2 and Si is a combination that allows etching reactions to proceed spontaneously. Such a combination is uncommon, and such gases are very reactive, unstable, and often hazardous. Most of the common gas molecules are more or less stable, which is obvious as they are not lone atoms. In order to perform etching efficiently, a gas is decomposed to a plasma, reactive radicals are generated, and the radicals are used to proceed etching reactions. This method is called plasma etching or plasma chemical etching and is widely used for dry etching.[5]

RF plasma discharge is commonly used plasma chemical etching. The basic setup is identical to the etching reactor shown in Section 3.5.2 (p. 83), and the parallel-plate plasma discharge apparatus shown in Fig. 3.14 is very popularly used. The barrel reactor shown in Fig. 6.8 is also popular. It is a capacitive coupled plasma reactor in which the opposite electrodes are arranged to surround a quartz bell jar.

Figure 6.9 shows schematics of plasma dry etching. One etching reaction proceeds in the following order:

> (i)   Decomposition of etchant gases and formation of radicals (p. 72)
> (ii) Diffusion and adsorption of radicals to surface
> (iii) Chemical reactions between adsorbed species and surface substance
> (iv) Desorption and vaporization of reaction products and exhaust

---

[5]In a more limited sense, plasma etching refers to the etching using a plasma and sometimes limitedly to this plasma chemical etching.

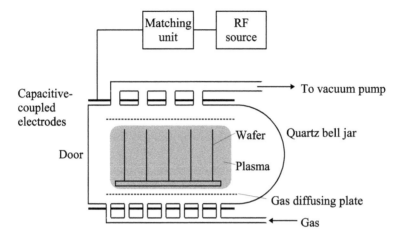

Figure 6.8 Barrel plasma reactor.

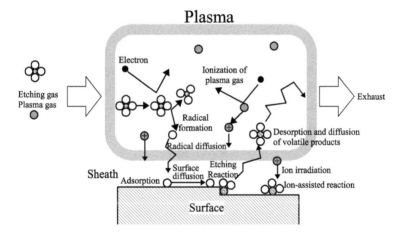

Figure 6.9 Mechanisms of plasma dry etching.

Actual chemical processes are very complicated, and the status of the etching plasma and reaction mechanisms vary depending on the etching gases, to-be-etched workpiece, the presence of photoresist, etc., whereas the above simple scheme explains a lot of crucial aspects of etching mechanisms. In Fig. 6.9, ion-assisted etching is also figured and its details are discussed in Section 6.6.1.

One important advantage of plasma dry etching is good time controllability. Etching starts immediately when the plasma is turned on and stops when turned off. As the etching is chemistry-based, the etching is more dependent on temperature than in physical etching. In addition to these, pressure and discharge power are also controllable. Many control parameters allow to tune plasma conditions precisely for etch patterning. In plasma dry etching described here, etching proceeds isotopically, as the etching species are diffusion-transported.

Next, let us see the mechanisms of plasma etching using F-based chemistry.

6.5.2.1 Si etching under F-based chemistry

CF_4[6] is a stable molecule and decomposes in a plasma, for instance,

$$CF_4 + e \rightarrow CF_3 + F, \tag{6.6}$$

resulting in the emission of F atoms and CF_3 and CF_2 radicals. They are neutral species and are not influenced by the electric field in a sheath, and diffuse and adsorb on the surface. F atoms are extremely reactive and react with Si spontaneously. The surface reaction occurring on the surface can be expressed, simply speaking, as[7]

$$Si + 4F \rightarrow SiF_4 \tag{6.7}$$

Referring to Table 6.3, the bond strengths of Si-F, C-F, and Si-Si are 6.2, 5.0, 2.4 eV, respectively, and Si-F > C-F ≫ Si-Si in order. That is, in a system containing C and F, the Si–F bond is most stable. However, the C–F bond is more stable than Si–Si, it is understood that a spontaneous reaction hardly occurs when CF_4 hits Si. By generating reactive and naked F atoms' plasma, they easily break the Si–Si bonds and form the Si–F bonds.

---

[6] or C_2F_6 or SF_6.

[7] Actual surface elementary reactions are thought to proceed as follows. Two fluorine atoms break a surface Si–Si bond and form SiF_2. SiF_2 is volatile but does not desorb promptly because the other two Si bonds are not broken. Next two F atoms react with SiF_2 successively, and the formed SiF_4, which is stable and highly volatile, desorbs from the surface.

Table 6.3 Chemical bond strengths (eV)

C–C	3.7	Si–Si	2.4
C–Cl	3.4	Si–Cl	4.2
C–F	5.0	Si–F	6.2
C=O	8.4	Si–O	4.6

The flux of molecules incident on the surface from the vacuum is [from equation (2.24)]

$$\sqrt{\frac{1}{2\pi mkT}}\, p \quad (\text{atoms/m}^2/\text{s}) \tag{6.8}$$

5–10% of these molecules dissociate to form radicals. Assuming that the etching reaction follows equation (6.7) and substituting the flux of the radicals, we can obtain the etching rate of Si (Problem 6).

As we can see in the periodic table that O and F belong to the same group, they have similar properties. The Si–O bond strength has a high value of 4.6 eV. As this value is slightly larger than that of Si–F, reactions between SiO_2 and F atoms do not proceed well as in the case of Si. This difference is beneficially used in selective etching of Si, for instance, for selective etching of Si using SiO_2 as an etching mask and for the selective removal of polycrystalline Si (poly-Si) on SiO_2. In experiments that irradiated Si with F atoms accelerated at low energies, it was observed that the etching selectivity of Si to SiO_2 was larger than 50 and it increased as the temperature decreased.

6.5.2.2 Etching of SiO_2 and Si_3N_4 using fluorinated gases

We have studied that Si can be etched well by using CF_4 and SF_6 plasmas; F atoms react well with Si but not with SiO_2 in principle. As a lot of F atoms are generated either in a CF_4 or SF_6 plasma, SiO_2 is hardly etched, whereas in the case of a CF_4 plasma, SiO_2 is etched as well as Si. This is because CF_3 radicals in the plasma are involved in SiO_2 etching.

It is known that, during etching, the SiO_2 surface is covered with a CF_x polymer layer formed through polymerization of radicals such as CF_3. This layer is less than 10 Å in thickness. SiO_2 is "eroded" with the CF_x polymer layer, not by the direct reaction with the incident radicals but by the generated volatile species such as COF_2, CO, and CO_2. These chemical processes can be written nominally as

$$SiO_2 + CF_x = SiF_4 + \left.\begin{array}{l} COF_2 \\ CO \\ CO_2 \end{array}\right\}$$

The CF_3 radicals are continuously supplied from the plasma, the polymer formation and erosion also proceed continuously, and the polymer thickness is maintained constant during etching.

Silicon nitride (Si_3N_4) can also be etched using fluorinated gases. Speaking from a chemical point of view, silicon nitride is similar to both Si and SiO_2, and its etching characteristics show intermediate behaviors.

6.5.2.3 Selective etching Si and SiO_2

Eventually, a CF_4 plasma etches either Si or SiO_2. In actual fabrication of microstructures such as electronic devices and MEMSs, the etch removal and patterning of an SiO_2 layer on an Si substrate and the selective etching of poly-Si on SiO_2 or Si are common processes.

It is known that the etching rate of Si or SiO_2 increases significantly when a small amount of O_2 is added to CF_4. This is because O_2 reacts with C atoms of CF_4 and forms stable CO_2, which results in intensive emission of excess F atoms.

$$\left.\begin{array}{l} O \\ O_2 \end{array}\right\} + CF_x \rightarrow \left.\begin{array}{l} COF_2 \\ CO \\ CO \end{array}\right\} + \left.\begin{array}{l} F \\ F_2 \end{array}\right\}$$

When O_2 is added at a concentration of 10–15%, the etching rate of Si becomes times higher, whereas that of SiO_2 does not increases significantly. This is used for selective etching of Si against SiO_2.

On the other hand, an addition of H_2 to CF_4 is known to decrease the Si etching rate and also improve the selectivity of SiO_2 dramatically. The addition of H_2 promotes polymerization and the formation of a thick Teflon[TM]-like fluorocarbon film (CH_xF_y polymer) that protects the Si surface against the incident radials and ions.

Si_3N_4 behaves as an intermediate chemical between Si and SiO_2 against F so that the selective etching against either Si or SiO_2 is not

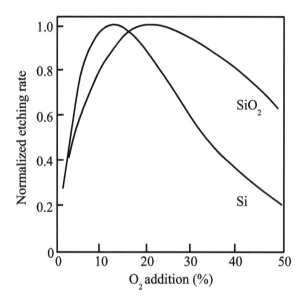

Figure 6.10 Effect of O_2 addition on etching rate of SiO_2 under CF_4 chemistry.

straightforward. When Si is etched selectively against SiO_2, an F-rich chemistry for Si etching is used, and when SiO_2 is selectively etched against Si, an F-deficient chemistry for SiO_2 etching is used.

Coffee Break: Heavy Load

In chemical etching, the size of the etching area sometimes affects the etching rate. This phenomenon is called loading effect. Usually the etching rate decreases as the area increases. The loading effect occurs when the etching amount is so large that the supply of the etching species becomes deficient.

The loading effect can occur locally within a wafer or widely within a reactor. In the former case, the etching rate is affected by, for instance, the area of an etching mask. In the latter, the etching rate varies from wafer to wafer in a multi-wafer batch reactor.

6.5.3 Photo-Assisted Chemical Etching and Electric Charging

Optical energy is large enough to induce chemical reactions, especially in short-wavelength light rays, such as ultraviolet. Chemical reactions induced by light are called photochemical reactions. Radicals can be photochemically generated and used to start surface etching reactions.

Photo-induced reactions are employed, for instance, to generate F and Cl radicals from Cl_2, ICl, and XeF_2 adsorbed on a substrate by a YAG laser so as to etch the substrate. A CO_2 laser can used to decompose SF_6 so as to generate F radicals for etching.

A higher energy (shorter wavelength) light can also be used. After adsorbing gas molecules such as Cl_2, an excimer laser light or a synchrotron radiation light is emitted. Photo-induced reactions take place selectively only in the light-emitted parts.

In plasma etching, the surface electric charging due to the ions or electrons in the plasma is of serious concern because the electric field induced by the accumulated charge affects the motion of the ions or electrons. This phenomenon hardly occurs in photo-induced etching, which makes this technique an attractive alternative.[8]

6.6 Physical and Chemical Dry Etching

The plasma chemical etching is a chemical etching technique that uses radicals generated in a plasma. Chemical etching is isotropic etching although it offers a very high etching rate. On the contrary, in physical etching (Section 6.4), the surface of a workpiece is irradiated with ions of an inert gas so as to sputter the surface atoms.

---

[8]The charged species, or the ions and electrons, incident on the surface pile up at the surface when they cannot escape electrically. This electrical charging occurs especially on insulative layers such as photoresist and oxides or even on a conductive layer that is electrically floated on an oxide. When the surface is electrically charged, the electrostatic field induced by the charges changes the motion of the incident ions, which deteriorates the preciseness of the etched patterns. In devices with integrated circuits, the piled-up charges can generate an electric field across the device (especially across a gate oxide layer) large enough to cause an electrical break down.

Physical etching is basically anisotropic etching but as the etching rate is much smaller than that of chemical etching, it is hardly used in practical or industrial micro-/nanofabrications.

By using a moderately reactive gas as the sputter gas, etching reactions are significantly promoted due to damage generation or energy transfer by **ion impact**, which enables to perform an etching that possesses both advantages of high etching rate of chemical etching and anisotropic etching of physical etching. This is the basic principle of "physical and chemical" dry etching, and from this, it is often called **ion-assisted etching**.

6.6.1 Reactive-Ion Etching

6.6.1.1 Principle and apparatus of RIE

A simple and convenient way to irradiate the workpiece (wafer) surface with ions is to bias the workpiece electrically negatively against the plasma. This method is called **reactive-ion etching**, abbreviated as **RIE**.

The simplest way to apply a negative bias is to use the self-bias (see Section 3.5.2.2), and the setup is almost identical to the RF sputtering apparatus (Section 4.3.2.1) but the sputtering target is replaced by the workpiece (wafer/substrate). The bias voltage is usually higher than a few hundred volts but does not exceed 2 kV; the nuclear stopping power is high and the sputtering yield becomes high in this preferable range of ion energy.

6.6.1.2 Reaction mechanisms

Let us see the principle of etching of the RIE, choosing Al and Si as examples.

Cl and F are congeners and either $SiCl_4$ or SiF_4 is volatile, but Si etching reactions by Cl atoms are not as fast as that by F atoms (Si–Cl bond strength is 4.2 eV). The etching reaction can take place when the ions collide with the Cl-adsorbing surface, knocking the Cl atoms beneath the surface, reducing the Si–Si bond distance, and breaking the Si–Si bond further. As a result, reactions between Si and Cl increase. Electron generation by the ion impact

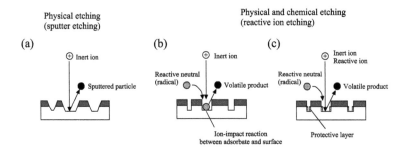

Figure 6.11 Principle of anisotropic dry etching.

and associated charge transfer also enhance the reactions.[9] As we have seen here, ion assist is expected to occur effectively in a system where the etching reactions that form volatile species do not occur thermochemically. When reactions occur spontaneously, like Si chemical etching with adsorbed F, the ion-assist effect is not very significant.

From the etching principles described above it is understood that

> In ion-assisted etching such as RIE, the etching reactions occur preferentially at the area where ions are irradiated.

As the ions are incident perpendicularly on the surface, the etching does not proceed laterally, leading to a very anisotropic etch topography.

Figure 6.11 is the one that supports the part of anisotropic dry etching shown in Fig. 6.1. (a) shows the sputtering etching described in Section 6.4 and (b) shows the principle of RIE.

A pure Cl_2 gas is rarely used as the source of Cl, instead chlorides such as CCl_4, BCl_3, and $SiCl_4$ are used. As they are heavy molecules, the ion-assist functions effectively and their radicals absorb excess free radicals that may contribute to spontaneous Si etching.

---

[9]The reactions between Cl and Si occur thorough electron transfer from the substrate Si to the adsorbed Cl atoms (Si is oxidized). Electron transfer occurs better in electron-rich n-type Si, poly-Si, and metals such as aluminum rather than in p-type Si or intrinsic Si, which can deteriorate the etching anisotropy.

Al is a material that dry-etches easily and has useful properties such as high anticorrosion property, high reflectivity, and high electric conductivity. For these reasons, Al is used popularly as an interconnect of LSIs and microcomponents of MEMSs.

Chlorine-based chemistry is also used for Al etching, and in the case of Al, the presence of the surface-native oxide (Al_2O_3) becomes a concern. This native oxide layer is very stiff and chemically stable and protects the bulk aluminum beneath the layer, which makes Al a good anti-corrosive material. Ion assist is very effective in removing the Al oxide semi-physically. To do this, heavier gas molecules such as BCl_3 instead of Cl_2 are used. Aluminum chloride ($AlCl_3$) is volatile, and therefore, once the oxide layer is removed, the etching of bulk Al proceeds well. Al–Si alloys and Al–Cu–(Si) alloys have been used as interconnect materials of Si-based electronic devices. The solute Cu can form a non-volatile residue as Cu chlorides have low vapor pressure.

Coffee Break: Damascene but Not for a Sword

Damascene has several meanings. Damascene steel was a very tough steel developed in Near East. Damascening refers to the inlaying technique of a metal.

Halides of transition metals have low volatility. Especially, the dry etching of noble and near-noble metals, such as Au, Pt, Pd, and Cu, is known to be difficult. When replacement of Al interconnects of LSI with Cu—that has a lower resistivity and thus improves device's performance—was attempted, dry etching of Cu became a serious and difficult challenge. Finally, researchers gave up and decided to fill pre-engraved trenches or holes with electroplated Cu. The excess Cu was polished away, which is identical to the inlayer technique developed in ancient days!! This technique came to be known as the Damascene method.

Do not worry about dry etching. It is still necessary to etch the trenches or holes instead of removing metals directly.

In RIE, high-energy (a few hundred eV–1 keV) particles are incident on the surface. This causes physical damage to the surface.

The thickness of this damaged layer is usually as thick as a few tens Å. This layer is removed by subsequent wet etching and annealing.

6.6.1.3 Anisotropic etching by sidewall protection

Very good anisotropic etching is needed for precision patterning. Side etching cannot be eliminated even in RIE due to obliquely incident ions and diffusion radicals. If the sidewall is covered with an inert protective layer, side etching can be suppressed and the etching anisotropy will improve. In RIE, this protective layer can be formed by tuning etching conditions, and good etching anisotropic etching profiles are realized. This is called **sidewall protection effect** [Fig. 6.11(c)].

The sidewall protection effect is realized by utilizing the deposition of a non-volatile film formed through plasma polymerization. Plasma polymerization is a process in which a polymer, usually a plastic-like organic substance, is formed due to the reactions between radicals in a plasma.

Polymer films are presumed to deposit uniformly over a surface, however, the bottom of a concave feature and the top flat surface

Table 6.4 Basic plasma etching gases and etching mechanisms

Etched materials	Etching gases	Co-added gases	Mechanism
Si, poly-Si, SiO$_2$, Si$_3$N$_4$	CF$_4$, C$_2$F$_6$, SF$_6$ NF$_3$, ClF$_3$, F$_2$	O$_2$ —	Chemical (radical reaction)
Al	CCl$_4$, BCl$_3$	(O$_2$)	Ion-assisted reaction
Mo, W, Nb	CF$_4$, SF$_6$	O$_2$	Ion-assisted reaction
Ti	CCl$_4$, CF$_4$	O$_2$	Ion-assisted reaction
Ta, TaN	CF$_4$	O$_2$	Chemical (radical reaction)
TiSi$_2$, TaSi$_2$, MoSi$_2$, WSi$_2$	CF$_4$, SF$_6$	O$_2$	Ion-assisted reaction
GaAs	BCl$_3$, Cl$_2$		Ion-assisted reaction
InP	CH$_4$, Cl$_2$		Ion-assisted reaction
Photoresist, organic substances	O$_2$	(N$_2$)	Chemical (radical reaction)

are sputtered continuously by the ions incident perpendicularly, therefore polymer films hardly deposit on these parts and the etching reactions proceed on the uncovered surfaces. Contrarily, sidewalls are not irradiated with ions and a thick (approximately 10–20 Å) polymer film deposits and protects the sidewall.

Even in the case of spontaneous relations such as Si etching through fluorine chemistry, the sidewall protection functions effectively, where the polymer layer protect the attack of F radicals and anisotropic etching is performed.

In order to enhance the polymerization, a small amount of H_2 is added, or a better polymerization gas than CF_4 such as C_2F_6 is used. A photoresist is an organic substance and functions as a source of hydrocarbons, which enhance polymerization. Other gases such as $C_xH_yF_z$ are also used.

6.6.2 Other Types of Ion-Assisted Etching Techniques

6.6.2.1 Reactive-ion beam etching

In usual RIE, a workpiece/wafer is placed on a negatively biased electrode so as to be exposed to a plasma. As the electric field in the plasma does not influence the motion of neutral radicals, the radicals diffused from the plasma to the surface induce isotropic etching.

The diffusion of radicals can be suppressed when the ion source (plasma) is placed far from the workpiece. The ions are extracted by the acceleration electrode(s) and introduced as a beam on the workpiece. The substrate bias is not needed. This arrangement is identical to the ion beam etching (ion milling) apparatus stated in Section 6.4 but a reactive gas is used in this case, therefore, this technique is called reactive-ion beam etching. The use of the reactive gas increases the etching rate compared to the case of inert gases and an improvement in etching selectivity is expected. Generally, the flux of incident ions is kept smaller than that of RIE so that the etching rate is smaller. For these reasons, this technique is used as an option when operating the ion beam etcher.

Reactive gases can be used in FIB. A reactive gas is fed near the portion bombarded by the ions, and the impact of ion bombardment accelerates the etching of the surface where the reactive gas is adsorbed. This is called reactive FIB or maskless ion-assist etching. The latter is because this is a successive pattering technique. The benefits of the use of reactive gases are the improvement of etching rate and the suppression of re-deposition of sputtered material.

6.6.2.2 Remote plasma etching

A parallel plate RIE (Section 6.6.1) uses self-bias in order to apply a negative bias to the workpiece. The bias voltage depends on the amplitude of the RF voltage or the electric power of plasma discharge. On the other hand, the gas decomposition and the formation of radicals and ions depends on the plasma density and electron energy (electron temperature). Therefore, when the discharge power is increased to enhance radical/ion generation, the bias voltage increases at the same time. This is not always favorable and therefore it is better to control radical/ion generation and bias voltage independently.

Another construction that may be an intermediate between the RIE and the reactive-ion beam etching is that a plasma is formed apart from a workpiece using electrodeless discharge and the ions are extracted to the workpiece by negatively biasing them through the use of another independent electric power source. This is called a remote plasma method. This construction allows independent control of the conditions of internal plasma and the energy of the ions incident to the workpiece. Recent advanced etching apparatuses, such as electron cyclotron resonance (ECR) plasma etchers, magnetron plasma etchers, and inductively coupled plasma (ICP) etchers, are high-density plasma etchers and employ this construction (see Section 3.5.3).

The etching rate depends on the incident flux of radicals and ions to the workpiece as well as the ion energy. The incident flux is mainly determined by the concentration of radicals and ions in a plasma. The higher the ion energy, the higher the etching rate; however, too high ion energy induces structural and electrical defects at the surface and deteriorates the etch-profile controllability due to too high sputtering effect compared to chemical effect. Remote plasma etching allows high-rate and high-precision etching as it enables independent control of plasma discharge and substrate bias.

The term 'remote plasma' is used for plasma techniques that do not use the substrate/workpiece bias. The bombardment by ions from the plasma is suppressed (the ion incidence due to the intrinsic voltage difference between the surface and the plasma—sheath potential—cannot be eliminated but it is much gentler than the intentional negative biasing), which allows minimizing the physical effects to etching. This method is called chemical dry etching (CDE) in a narrow sense, and is sometimes called after-glow etching, although this definition is somewhat less rigorous.

Coffee Break: Etching Endpoint Detection

As the etching selectivity cannot be infinite, the underlayer of a to-be-etched film can get etched when the etching time is too long.

The structures processed in dry etching are usually very small and visual observation of the etching endpoint is quite difficult. The workpiece should not be taken out from the process chamber; it is not a very effective way even if it is possible. For these reasons, methods of etching endpoint detection haven been developed.

Atoms and molecules in a plasma emit light during the relaxation processes. The wavelengths of emission are specific to and different for atoms and molecules (emission lines). The etching endpoint is determined by measuring the intensities of the emission lines of atoms/molecules of etched substances. At an endpoint, the emission from the etched layer disappears, and at the same time, the emission from the underlayer will appear.

┌─────── **Coffee Break: Words Used for Scaling and Counting** ───────┐

The unit used to measure length is "meter." Likewise, kg is used for mass and s for time. They are uniquely used in sciences as "the SI units" and are used daily in most of the countries. In some countries, different units such as inch and pound are used in daily life.

Metric prefixes are, for example, centi, kilo, micro, and nano, which correspond to 10^{-2}, 10^3, 10^{-6}, and 10^{-9}, respectively. These prefixes are used with units, such as nanometer (nm).

Metric prefixes basically increase or decrease every 1000 times. This is based on the short-scale numerals in western countries. In different countries, especially in Asian countries, different numerals are used. The next of one (10^0) is ten, the next of ten (10^1) is hundred (10^2), then, thousand (10^3), "ten-thousand" (10^4), ten "ten-thousand" ($10\ 10^4$, not in plural), hundred "ten-thousand" ($100\ 10^4$), ..., "hundred-million" (10^8), ten "hundred-million" ($10\ 10^8$), and so on.

In Asian countries, "counter words" are used with numbers. A counter word is placed after a number, in the same manner as units. Different counter words are used for counting different objects. For people, pencils, automobiles, ships, fishes, molecules, etc., all counter words are different. Although these counter words are recognized and used identically as units, even in scientific descriptions, they never appear in SI units—they are just non-dimensional.

└──┘

Problems

(1) What causes side etching?
(2) Describe the principles and methods for Si anisotropic wet etching.
(3) A 1 µm thick Cu film is etched with Ar ion beams accelerated at 0.6 keV. W is used as an etching mask. Estimate the thickness of the W needed. The ion beam is incident normal to the specimen.

(4) For what reason is an FIB apparatus equipped with a mass-selector.

(5) Describe the mechanism of plasma etching. Clarify key factors of plasma etching.

(6) Calculate the Si etching rate in a CF_4 plasma at 0.5 Pa. The temperature of the plasma gas is 500 K and the F concentration is 5%.

(7) Describe the dry etching mechanism of Si and SiO_2 in a CF_4 plasma. Describe how to etch Si and SiO_2 selectively with each other.

(8) List isotropic and anisotropic dry etching methods and describe their physical and/or chemical principles briefly.

(9) Describe reactive-ion etching (RIE).

(10) Describe similarities and differences between an RF reactive sputtering apparatus (Fig. 4.9) and an RIE apparatus.

(11) Describe the principles of the ion-assist etching effect.

(12) Anisotropic etching of RIE is based on two different principles. Describe each citing their similarities and differences.

(13) In RIE, vertical sidewall etching can be performed without ion-assist effects or sidewall protection when sputter etching and chemical etching occur simultaneously. Why?

(14) Describe loading effect.

Chapter 7

Photolithography

7.1 Introduction

Modern microelectronic devices such as integrated circuits (ICs) consist of microscale components assembled in a complicated manner. Each component is produced through a series of processes of, such as, thin film deposition (additive process), deposition of etching mask materials (additive) and their pattering (subtractive), and etching of the thin films (subtractive). This sequence is repeated using different film materials and a stack of different components is assembled finally.

In the previous chapters, we learned the principles of each process. In this chapter, we will study about the technology used to pattern etching mask materials using light. This process is called photo fabrication, in general, but, is also called photolithography much more popularly in microelectronics. Photolithography is a process to "print" copies using a photomask that corresponds to a master through a sort of photomechanical technology. This technique has a very high productivity and enables a precise pattern transfer and position alignment, which is ideal for mass-production of micro- and precision patterns.

Micro- and Nanofabrication for Beginners
Eiichi Kondoh
Copyright © 2021 Jenny Stanford Publishing Pte. Ltd.
ISBN 978-981-4877-09-1 (Hardcover), 978-1-003-11993-7 (eBook)
www.jennystanford.com

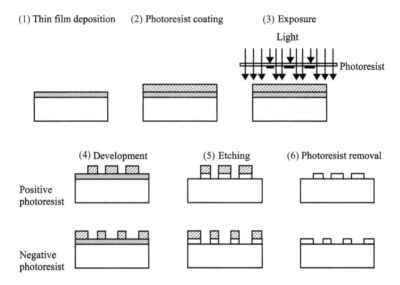

This book explains the principles of micro- and nanofabrication using gas as a "tool". Photolithography is completely different from this, but it is an essential technology in micro- and nanofabrication. For this reason, this chapter is devoted to this technology. As the chemical and physical aspects and backgrounds are completely different from those discussed in previous chapters, the basic concepts and principles are being discussed here.

7.2 Introduction to Photolithography

In photolithography, (i) a thin film to be patterned is deposited; (ii) a photosensitive organic material called photoresist is coated and then dried; (iii) the photoresist is exposed to light through a mask pattern called photomask or reticle; (iv) the exposed (or unexposed) part is removed the development, where the exposed part is removed when a positive photoresist is used whereas the unexposed part is removed when a negative photoresist is used; (v) the thin film is etched using the patterned photoresist as an etching mask; and finally, (vi) the remaining photoresist is removed.

Figure 7.1 shows an outline of photolithography.

(1) thin film deposition → (2) photoresist coating and baking → (3) exposure → (4) development → (5) etching → (6) photoresist removal

By using an optical technology, the geometric pattern on the photomask is transferred to the photoresist on a substrate and is finally transferred to the film on the substrate. Figure 7.1 shows a cross section. A bird's-eye view of the process steps of the fabrication of semiconductor integrated circuits is shown in Fig. 7.1, which shows the manner in which a shadow pattern of the light that passes through a photomask is reduced by lenses and projected on the wafer.

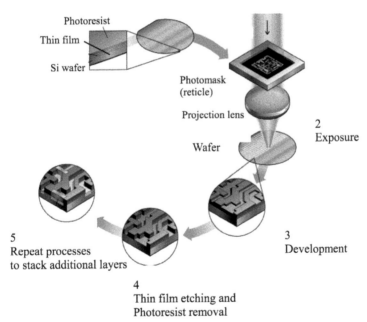

Figure 7.1 Lithography process for manufacturing of semiconductor integrated circuits. Adapted from Scientific American.

Coffee Break: Lithograph

Lithography is a printing technique using a plate of a stone as a block. It was invented in Germany in the 18th century. After drawing on the plate with an oil-based material, the plate is dipped in water to make the surface water-wettable. Then this hydrophilic part repels an oil-based ink for printing, and the ink attached on the drawing is printed on paper. This technique became very popular because it does not require engraving, is printable as drawn, and can express intermediate brushwork.

The plate is durable enough to print several copies. Lithography is also used as an industrial multicolor printing technique. Since no engraving is required, it allows to print a photograph by using a photosensitive material. This is photolithography. So, photography is a technique to print photographs.

Lithograph poster (Alphonese Mucha, 1897)

Using the analogy of "reprinting" of a picture in which a photomask is the original plate that corresponds to a film and a wafer on which the photoresist is coated corresponds to a photographic paper.[1] Many extra copies are made from one film, and likewise, the same pattern is copied to many wafers using one photomask. As the wavelength of light is very short, the use of optical technology allows to obtain high resolution easily. Photolithography is the best method for the mass production of small patterns.

In the production of components of very complicated and multifunctional devices such as integrated circuits, a lot of layers such as the internal passivation layer, electrodes and interconnects, and insulation layers are stacked. The process steps from (i) to (vi) stated above are repeated. A different photomask is used in each step, and additional photomasks for the fabrication of a sacrificial pattern may be sometimes used in intermediate steps.

Misregistration must not be allowed in multicolored printing. Likewise, an upper film layer must be patterned on the underlying film pattern without an alignment error. The sizes of internal components of integrated circuit chips range from a few μm to 10 nm, so that extremely advanced precision technologies are required to ensure perfect alignments over many layers. For this reason, the cost of the photolithography is a very high. For instance, 15–30 photomasks are used for the main memories of personal computers, called dynamic random-access memories (DRAMs), and this makes the lithography cost about 1/3 of the total cost.

7.3 Photoresist Process

Photoresist is a sort of photosensitive organic material. There are two types of photoresists—positive type and negative type. The exposed portion of the **positive photoresist** dissolves in the development solution and the unexposed portion remains after the

---

[1]This analogy is getting less comprehensive as digital cameras have become so popular.

development. For the **negative photoresist**, the remaining pattern is opposite to this.

A photoresist is a viscous liquid coated as a very thin layer on a wafer. After the coating procedure, the photoresist is heated to 80–100 °C to be hardened (called prebake). The photoresist changes its chemical properties when exposed to light. For the positive photoresist, the exposed part becomes soluble to a developer, and for the negative photoresist, the exposed part becomes insoluble to a developer. **Development** is a process to develop an image on the exposed photoresist in an aqueous solution or organic solvent. The image after the development becomes identical to the image on the photomask for the positive photoresist, and vice versa, for the negative photoresist (see Fig. 7.1). The developed photoresist is further heated to harden to make it resistive enough against the following processes.

7.3.1 Photoresist Materials

A positive photoresist is a mixture of a resin soluble to a developer and an insoluble photosensitizing agent. This mixture is dissolved in an organic solvent that vaporizes during prebaking conducted after the photoresist coating. The prebaked photoresist is insoluble in the developer but the photosensitizing agent becomes soluble after exposure.

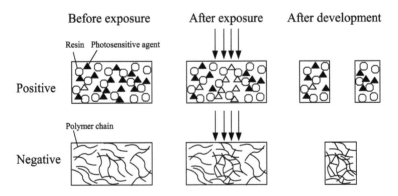

Figure 7.2 Structure changes upon photochemical reactions for positive and negative photoresists.

The most common positive photoresist is a DQN resist. The same photosensitizing agent like a cyanotype is used,[2] DQN abbreviates from diazoquinone (DQ), a photosensitizing agent, and novolac (N) which is a matrix resin used as an adhesive of plywood. Diazoquinone converts to a ketone when exposed to light and to carbonic acid when reacted with water. The carbonic acid dissolves well in alkaline solutions. The developer for the DQN photoresist is alkaline. A mixture of xylene and acetate is used as a solvent.

The principle of the negative photoresist is very different from that of the positive photoresist. A sort of hardening reaction takes place upon exposure to light, during which the chains of polymer molecules react with each other forming bridges between the chains—so called cross-linking reactions; as a result, the photoresist becomes insoluble to a developer. Negative photoresists have a high photosensitive reaction and adhere well to a wafer. Due to the cross-linking, a negative photoresist shows better chemical and plasma tolerance, which makes it appropriate as an etching mask. Developers of negative photoresists are organic solvents and they tend to swell, reducing the pattern preciseness.

7.3.2 Photoresist Coating

A photoresist is coated onto a substrate or wafer by the **spin-coating** method. A liquid photoresist is applied to a spinning wafer that is vacuum-chucked to a pedestal (Fig. 7.3). A photoresist expands due to the centrifugal force while keeping a good thickness uniformity due to the wetness and viscosity of the liquid. The thickness t is known to be determined by the rotation speed ω (Fig. 7.4).

$$t \propto \frac{1}{\sqrt{\omega}} \tag{7.1}$$

A typical rotating speed is 2000–3000 rpm. The thickness depends on the viscosity and obviously, the lower the viscosity, the higher the thickness. An adhesion promoter, such as HMDS

---

[2]The cyanotype or blueprint is a photocopying technique used widely before electrophotography, currently known simply as photocopying, became popular. It is still used in graphic arts.

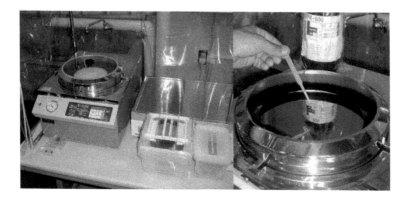

Figure 7.3 Laboratory spin coater (left) and photoresist coating work (right, conceptual).

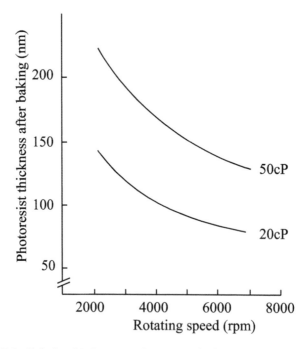

Figure 7.4 Relationship between photoresist thickness, rotation speed, and viscosity. Based on a catalog of Tokyo Ohka Kogyo Co.

(hexamethyldisilazane), is often applied to the wafer directly—HMDS is a liquid—or as a vapor, prior to the coating.

7.4 Photomask

As already stated, a photomask corresponds to a negative film in photograph copying. A negative film has an image consisting of transmissive and opaque (and half-tone) areas on a supporting plastic film. A photomask uses a transparent glass as a supporting material and has a component or circuit pattern consisting of completely transmissive and opaque diagrams.

7.4.1 Photomask Materials

Silica glass or soda-lime glass is used as a photomask plate. Silica glass is transparent from visible lights to ultraviolet rays and is chemically stable, whereas soda-lime glass is affordable and large sizes are available. Photomask glasses are very carefully fabricated and their specifications, such as chemical purity, transparency, thermal expansion coefficient, flatness, and surface roughness, are controlled at the extremely high level.

A raw, unpatterned photomask—called a (photo-)mask blank—is covered with a layer of an oblique matter. A sort of photolithography is used to pattern this layer; but before we explain this method, let us see the materials of the oblique layer.

Simple and traditional photomasks use either "emulsion" mask or "hard" mask. An emulsion is a gelatin layer in which silver halide (AgBr) is dispersed (emulsion). When exposed to a light, AgBr is reduced. The deposited Ag particles exhibit black opaque contrast, and the unexposed area disappears. The thickness of the emulsion layer is about a few μm, and the practical pattern resolution is the same. Chromium is used as an opaque layer of hard masks. Cr has a high light-adsorption coefficient, is chemically stable, and shows low internal stress. Hard masks are used to form finer patterns.

A emulsion mask is a sort of photographic plate, and can be readily used as a photomask after usual exposure and development procedures. For chromium masks, whole steps of photoresist coating, exposure, development, and chromium etching are needed to be carried out as in usual wafer processing.

Historically, the master patterns used for exposure in photomask fabrications were made by using a red-colored, semi-transparent masking paper (Rubylith). The red film is used because it does not transmit blue light or ultraviolet light, reducing projection exposure. The master patterns were also made with a pattern generator, which is a machine that makes unit graphical circuit patterns using an adjustable opening, where the patterns are used successively as masks for reduced projection exposure. In modern photomask fabrication, a scanning laser or electron beam is used to draw patterns directly on a mask blank. Both are successive processes. The pattern preciseness improves in this order. Electron-beam lithography is used only when the pattern size is less than 1 μm. Residual or deficient defects are corrected by laser abrasion removal or additive/removal operation by FIB. A numerous number of wafers are fabricated from one photomask. Therefore, the inspection and correction of photomasks are extremely important and very advanced technologies are used for them.

Particles and dusts on a photomask make unnecessary shadows and include the reduction of pattern preciseness and pattern defects. The cleanliness of photomasks should be controlled with utmost care. In projection printers described later, a cover called pericle is used to prevent contamination.

7.4.2 Designing Photomask Patterns

Computer-aided design (CAD) tools are used to design patterns of photomasks. Actual circuits and components are three-dimensional and composed of stacked-pattern layers. Patterns of each layer is drawn on the photomask of each layer. CAD systems used for photomask designing are usually linked to CAD systems for circuit design in the case of integrated circuits or other electric circuits.[3] The photomask CAD data is converted to the data for computer-

---

[3]Circuit functions and operations are described by a computer language called hardware description language (HDL).

Figure 7.5 A photomask being taken out from a container.

aided machining (CAM) data for lithography tools, such as a laser or electron-beam lithography tool and a pattern generator.

The dimensions of a circuit pattern on a photomask is determined by the specification of the exposure tool, usually same as the real scale or 1/5—1/10 of the real scale. Although one whole photomask can be used for one circuit, usually, the photomask area is sectioned into many grid square areas where the same circuit pattern is placed repeatedly. In this way, many integrated circuit chips are produced from one wafer, and more importantly, the production yield improves.[4] When the reduced projection exposure is used, the projection area is much smaller than the wafer, the exposure is repeated by changing the position of the exposure within the wafer. The number of circuit chips on the wafer is (number of repeat) × (number of chips on photomask). This is huge, which is an advantage of using the reduced projection.

---

[4]If the whole area is used, the whole wafer is to be abandoned when a small defect presents on the surface or photomask.

7.5 Exposure

7.5.1 Printers

When you remove, for example, a picture or postcard pinned on the wall, you might see its trace, as the wall color has been faded. The color fading is a photochemical reaction, and the shape of the card is transferred to the wall. In this sense, this is identical to photolithography, where a mask is contacted to the layer and the shape is transferred at the same size.

On the other hand, when a movie film is projected with a cinema projector in a theater, the image is magnified on the screen. If a photosensitive material is placed on the screen, the magnified image is exposed; the projection printer or enlarger in photography uses this principle. And, it can easily be reduced or same-sized images can be projected by exchanging lenses.

The enlargement is of no use in micro- and nanomanufacturing, therefore, reduced or same-sized projection is employed (projection printing). In addition to this, contact exposure (contact printing), in which the photomask is brought in direct contact with the photoresist, and proximity exposure (proximity printing), in which a small gap is left between the photomask and the photoresist, are used.

7.5.1.1 Contact and proximity printers

In contact printing, a photomask is directly contacted with the photoresist on a substrate. A **contact printer** has the simplest configuration among printers. Obviously, the image printing magnification is unity.

The mask layer of the photomask is faced to the wafer; if it is opposite, the contrast of the shadow (mask image) becomes weaker due to the increase in the light path length by the photomask thickness that induces diffraction and interference of the incoming light. The photomask is pressed on the wafer so as to ensure good overall contact of the wafer area, as the wafer and the photomask cannot be placed perfectly in parallel due to the wrap of the wafer and/or photomask and the deformation of the printer stage. Even so, when an additional layer is printed on an already-patterned underlayer, the printing is sometimes carried out without contacting the photomask. The resolution of contact printing is usually about 1 μm.

A direct contact between the photomask and the wafer is problematic because it easily causes defects or damages on either the photomask or the wafer. The attachment of particles and dusts easily occurs.

In **proximity printing**, a small gap of 10–50 μm is made between the photomask and the wafer. A gas such as nitrogen is flown to this gap and this gas functions as a "cushion" or "lubricant" so that the photomask can move on wafer smoothly. However, the photomask to wafer spacing is increased, the optical distance increases, requiring a parallel light ray source. Due to this gap, the pattern transfer resolution becomes worse than contact printing.

In contact and proximity printing, the wafer and photomask are contacted or closely placed each other. Caution is required in exchanging the photomask or the wafer, which limits the freedom in mechanical or spatial designing as well as the throughput.[5] For instance, for a contact printer such as the one shown in Section 7.5.1.2, the wafer stage is fixed, and therefore the photomask must be moved before and after each exposure. This causes an increase in the alignment of the photomask onto the patter on the wafer. In addition, the photomask and the wafer are in the same ambience, which is perhaps a more crucial issue. The cleanliness of the photomask should be more carefully controlled so that it can be placed in a different environment.

7.5.1.2 Projection printers and steppers

In **projection printing**, the photomask is fixed in a different environment than the wafer, apart from the wafer. This allows quick transfer and exchange of wafers. Projection printers are now widely used in microprocessing industries.

Projection printers form an image of the photomask onto a wafer using an optical projection system. The use of the projection

---

[5]Throughput means the amount or magnitude of objects or activities processed per unit time. In wafer processing, this means the number of wafers processed per unit time (hour, day, etc.).

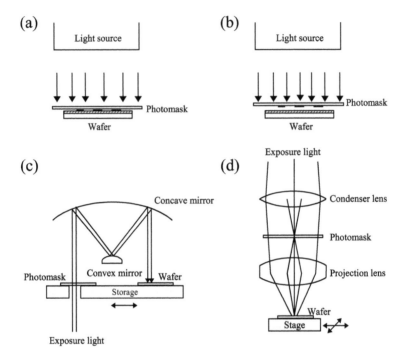

Figure 7.6 Photolithography printers (a) contact printer, (b) proximity printer, (c) (same-size) projection printer, (d) reduced projection printer (stepper).

system allows to form a high-resolution image even though the photomask is located far away from the wafer. The generation of photomask flaws can be completely suppressed and the photomask environment is controlled independently of the wafer environment.

The same-size projection printer [Fig. 7.6(c)] was developed in 1970s and became very common in semiconductor production.[6] A strip-shaped (actually arc-like to reduce aberration) light is illuminated onto the photomask, and the light image passed through the photomask is condensed by a concave mirror, then reflected, and then projected on the wafer by using another concave mirror. The light is illuminated onto a part of the photomask and the wafer; the

---

[6]This reflection optics is called the Offner system.

photomask and the wafer are moved simultaneously so as to overall scan the photomask and the mirror.

In such a projection system, the oblique image of the photomask is focused and projected onto the wafer. Particles and dusts, even if any, do not adhere directly on the photomask whereas they are present on a pellicle which protects the photomask so that their images are not focused on the wafer. In addition, the image scanning arrangement allows to project a part of the photomask, which improves the optical preciseness and thus pattern resolution. It is noted that the reflection system has a higher optical resolution and thus enables a large area exposure due to the smaller chromatic aberration as well as the ease of mechanical fabrication of mirrors; this is similar to the fact that reflection telescopes are better than transmission telescopes.

However, projection printers are not used practically. Reduced projection printers that have a so-called "step and repeat" function are used commonly in semiconductor industries. This type of printer is usually called a stepper [Fig. 7.6(d)]. In the step and repeat function, the wafer stage is mechanically moved with repeating exposures at regular intervals. The reduction ratio is usually 1/5 but can also be 1/10. As the photomask image is reduced, the exposure is repeated at regular steps, which is the reason behind the origin of the name, so as to expose the overall surface (see Fig. 7.1).

Compared to the same-size projection printer, the stepper has the following advantages.

(i) Patterns of the photomask are reduced, allowing further miniaturization.

(ii) Photomasks can be smaller than the wafer, which comforts photomask production. (In same-size projection printing, the photomask should be as large as the wafer so as to minimize wafer area penalty.)

(iii) Photomask defects are reduced, leading to improved precision.

(iv) Area of exposure is small and the exposure conditions can be controlled at each shot, which leads to a good in-plane uniformity.

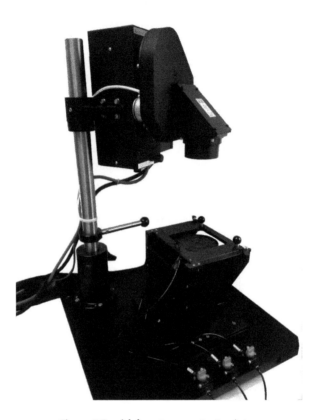

Figure 7.7 A laboratory contact printer.

Steppers are equipped with an automatic focusing system and an automatic alignment system, which contribute to realize high preciseness. However, repeating exposures many times take a lot of time per wafer, which reduces the throughput. The most advanced optical components, such as lenses and a light source, and a highly precise and complicated mechanics are used to realize nanometer-level resolution and alignment. The steppers are tremendously expensive.

In reduced projection printing, as shown in Fig. 7.6(d), the light passed through a condenser lens is illuminated on the photomask, and the image of the photomask is reduced by a projector lens and is formed onto the wafer. The performance of the lens system is expressed by **numerical aperture, NA**.

Figure 7.8 ArF stepper (scanner) (ASML http://www.asml.com).

$$\mathrm{NA} = n \sin \alpha \qquad\qquad (7.2)$$

where α is the half value of the angular aperture, n is the refractive index of the medium present between the lens and wafer. Air and nitrogen are common media ($n = 1.0$) and water ($n = 1.33$) is also used in advanced high-resolution lithography. A typical NA value range is 0.16–0.5.

The **resolution** W is determined by the performance of the focusing lens and is given by the following equation:

$$W = k\frac{\lambda}{\mathrm{NA}} \qquad\qquad (7.3)$$

where k is a constant. k is typically about 0.75 and its theoretical value is 0.61. The resolution improves as NA increases, whereas the depth of focus (DOF)

$$\frac{\lambda}{\mathrm{NA}^2} \qquad\qquad (7.4)$$

becomes smaller. Processed wafers are always wrapped due to the stresses caused by deposited thin films, the circuit patterns on the wafer make a bumpy non-leveled topography. To assure a good focus and pattern resolution globally (in-plane) and locally, a smaller DOF is preferred.

To make the resolution smaller, it is straightforward to use a shorter wavelength light. However, in general, the adsorption coefficient increases as the wavelength becomes smaller, which makes it difficult to fabricate bright large-diameter lenses.

In projection printing, diffraction due to the lens–pupil effect is an issue. An ideal lens can form an infinitesimally small point image at a focal point, condensing parallel incident light to an infinitely large aperture. Actual lenses have a finite area and function as a pupil placed in the light path. The incident light is diffracted at the edge of the pupil, and the diffracted light distorts the other ideally propagating rays. As a result, the focused image becomes a blurred circle[7] that cannot be distinguished from the image of a same-size circle projected by an ideal lens. The theoretical resolution mentioned above refers to the size of this circle. As stated, the resolution improves when NA becomes larger; the reason for this is that a lens with larger NA has a larger identical aperture.

When sizes of opaque patterns on the photomask are as small as the light wavelength, these patterns diffract the light. This is also a serious issue. Figure 7.9 shows the distribution of the intensity of the light that was passed through a slit opening. Figure 7.9(a) shows an ideal case, whereas when the light diffracted, it propagated behind the shadow of the opaque pattern as shown in (b). As a result, the blurred image is projected on the wafer. In addition, when the openings are periodically arranged, interference occurs, which also degrades the pattern definition. In advanced lithography, the photomask patterns are designed taking these effects into account. For instance, a sharp edge is projected as a blurred round image, the photomask pattern is designed to compensate the unnecessary light distribution or to intensify the brightness. Half-tone masks and complicated geographic patterns are used to realize this. This is a kind of inverse problem and is to be solved computationally.

---

[7] Fraunhofer diffraction

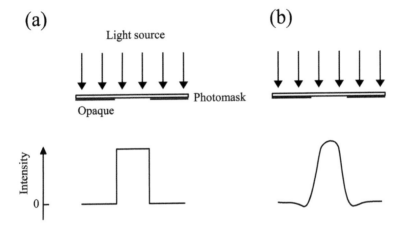

Figure 7.9 Diffraction in lithography.

Equation (7.3) tells that patterns smaller than the theoretical resolution cannot be fabricated. These issues can be averted in part by the method called multiple patterning (double, triple, etc.). In multiple patterning, one pattern definition is formed by dividing it into different exposures. For instance, in the simplest case, the exposure of narrow lines is achieved by overlapping two off-shifted images. Dielectric hardmasks are often used sacrificially to decrease the number of photomasks. As understood from its principle, multiple patterning can be limitedly used for, such as, periodical "lines and spaces patterns." As integrated circuits have many symmetrical, periodical, and rather simple structures, multiple patterning is very effectively used to form sub–50 nm features, although it is very expensive.[8]

The light incident on the wafer will be reflected by the surfaces and interfaces of a photoresist, leading to the **interference** inside the photoresist layer. As the phase shifts upon reflection, the overlapping of the incident light and the reflected light forms a standing wave within the layer, i.e., intensity distribution along the thickness direction. The photoresist always has a certain non-uniformity, the standing wave formation also has an in-plane non-uniformity, which results in an in-plane exposure non-uniformity.

---

[8] Frensel diffraction.

Concave–convex features of patterned structures can scatter or diffract the reflected light, which causes undesired exposure under the shadows of oblique mask patterns. To decrease the reflection coefficient of the surface or to use an anti-reflective coating is very effective to suppress these phenomena, especially for short-wavelength lights, as the reflectivity often increases for a shorter wavelength range, and for laser light sources, as laser light has good coherency and therefore induces interference easily.

7.5.2 Light Sources

Simple printers, such as contact and proximity printers, are equipped with a **mercury-vapor lamp**. The principle of light emission of these types of lamps is discharge or plasma generation of high-pressure mercury vapor. The gas pressure in the lamp bulb is usually about 1 atm when being cooled and not discharged, whereas the pressure increases to a few tens atm and the bulb surface temperature is said to be as high as 600 °C. The discharge occurs at a tip on an electrode which is about a few mm away from another electrode. The emission region is very limited so that the mercury-vapor lamp can be regarded as a point light source in terms of designing of optics. The emission voltage is a few kV and the electric power is 500–1000 W. The bulb is cooled to suppress the damage of the bulb glass and the sputter erosion of the electrodes.

The wavelength region of the light of the mercury-vapor lamp is widely distributed, whereas some strong emission occurs as particular wavelengths. Those emissions appear as lines in the spectrum and are therefore called **emission lines**.

The emission lines correspond to the relaxation processes of excited Hg atoms in the plasma (see Section 3.3.2) and each relaxation process has an emission line. Many emission lines can be seen in the spectrum of the mercury-vapor lamp, and a particular emission line is selected by using an optical filter. Intensive lines have historical labels; and among them, g-line (436 nm) and i-line (365 nm) have been popularly used in lithography (see Table 7.1).[9]

---

[9]The broad background spectrum is due to the radiation of high-temperature electrons (40,000 K). The electron gas behaves as a gray body. The shortest wavelength is 75 nm.

Table 7.1 Wavelengths of single wavelength light sources

Light source	Emission line/exciplex	Wavelength (nm)
Mercury vapor lamp	g	436
	i	365
Excimer laser	KrF	248
	ArF	193
EUV		13.5

In recent lithography for integrated circuit fabrication, **excimer lasers** are used.[10] Common excimer lasers utilize the emission from an exciplex (molecule) that is formed from a noble gas and a halogen gas.

$$Xe^* + C\ell_2 \rightarrow XeC\ell^* + C\ell \tag{7.5}$$

This excited molecule is very unstable and decomposes. Therefore, gas excitation is needed for it to repeat continuously in order to maintain the emission. Gas discharge at a voltage of 10–20 kV is commonly used. For these reasons, the emission occurs for a very short time like a pulse.

Excimer lasers have a much more complicated set-up than a simple mercury-vapor lamp. Formerly, one of the strong issues was the need of frequent maintenance; however, with technical progresses and improvements, practical issues have solved. Photoresists and optical components that endure short-wavelength lights have also developed. Advanced steppers are also called scanners, as the laser light beam is not expanded to illuminate the whole lens area but instead the lens is partially illuminated and is scanned in order to increase the power density. This motion is identical to the old same-size projection printer.

The extreme ultraviolet (EUV) lithography is the most advanced technology available currently. The wavelength of the EUV for lithography is 13.5 nm, much shorter than excimer laser lights. In fact, its wavelength is in the range of soft X-ray (0.5–15 nm or 0.1–2 keV), whereas it is called a sort of ultraviolet in the lithography. The light source of the EUV is the discharge of Sn vapor excited by CO_2

---

[10]An excimer stands for an excited dimer, exciplex to be precise.

laser. As the wavelength is too short, usual lenses and mirrors cannot be used; instead reflection from Mo and Si multilayers is utilized. The theoretical resolution is very short according to Eq. (7.3), whereas the actual resolution is not as small as this due to a lot of limitations of optics and materials.

Problems

(1) Figure 7.10 shows a virtual structure fabricated by conducting the deposition of polycrystalline silicon and aluminum thin films and patterning them sequentially. Design the photomasks for the cases of using positive photoresist and negative photoresist. Assume proper dimensions, if necessary.

(2) For the above structure (Fig. 7.10) draw A–A′ cross-sections for each process step.

(3) State why the particle/dust attachment onto a photomask is a serious issue.

(4) Calculate the theoretical resolution for a g-line shrinkage projection printer. Use $k = 0.75$ and NA $= 0.4$.

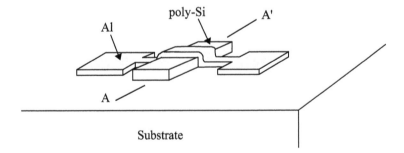

poly-Si   A′
Al

A

Substrate

Figure 7.10 Figure for Problem 1.

Appendix A

Symbols and Variables

For the readers' convenience, symbols, annotations, and variables
that appear in this book have been summarized. The same symbols
or variables may have different meanings in different sections. Units
of numbers and constants that appear in this book have been listed
in this table unless otherwise specified.

Table A.1   List of symbols and variables

Symbols	Definitions	Units	Symbols	Definitions	Units
a	lattice constant	m	α	acceleration	m/s^2
A	area	m^2	α	angle	rad
$B, \mathbf{B}$	magnetic density	T	ΔU	molecular internal energy	J
C	capacitance	F	ε_0	permittivity of vacuum	F/m
d	molecular diameter	m, Å	η	sputtering yield	–
D	characteristics length, thickness	m	η	lattice mismatch	–
D	irradiation energy density	J/m^2	θ	collision frequency	s^{-1}
E	energy	J	λ	mean free path	m

(*Continued*)

Micro- and Nanofabrication for Beginners
Eiichi Kondoh
Copyright © 2021 Jenny Stanford Publishing Pte. Ltd.
ISBN 978-981-4877-09-1 (Hardcover), 978-1-003-11993-7 (eBook)
www.jennystanford.com

Table A.1 (*Continued*)

Symbols	Definitions	Units	Symbols	Definitions	Units
$\mathcal{E}$	electric field	V/m	λ	wavelength	m
$f(v)$	distribution of velocity v	–	ν	light frequency	1/s
F	force	N	ρ	charge density	Cm^{-3}
g	gravitational acceleration	Jm/s^2	ρ	mass density	kg^{-3}
G	free energy	J/mol	σ	collision cross section	m^2
H	enthalpy	J/mol	ω	angular velocity	rad/s
$J, \mathbf{J}$	flux	$m^{-2}s^{-1}$	ω	rotating speed	s^{-1}
J	current density	Am^{-2}			
k	Boltzmann constant	J/K	x, y, z	axes	
ℓ	distance	m	e	electron	
m, M	mass	kg	n	neutral species	
M	molar number	–	i	ion	
M_w	molar mass	kg/mol			
n	density	m^{-3}			
n	refractive index	–			
N_A	Avogadro number	–			
$\overline{v}$	average speed	m/s			
$\overline{v^2}$	mean square speed	m^2/s^2			
$\overline{v}^2$	square of mean speed	m^2/s^2			
p	pressure	Pa			
p_T	total pressure	Pa			
P	probability	–			
Q	gas flow rate	$Pa·m^3/s$			
r	radius	m			
R	gas constant $(= kN_A)$	J/mol/K			
S	area	m^2			
S	pumping speed	m^3/s			
S_{eff}	effective pumping speed	m^3/s			
t	time	s			
T	absolute temperature	K			
$v, \mathbf{v}$	velocity	m/s			
v_m	most probable speed	m/s			
V	volume	m^3			
x, y, z	position	m			

Appendix B

Basic Physical Constants

Table B.1 Basic physical constants

Elementary charge (electron charge) e	$1.60217733 \times 10^{-19}$ C
Electron rest mass m_e	$9.1093897 \times 10^{-31}$ kg
Planck constant h	$6.6260755 \times 10^{-34}$ J·s
Avogadro constant N_A	6.0221367×10^{23} mol^{-1}
Molar volume of an ideal gas (0 °C, 1 atm)	2.241410×10^{-2} m^3/mol
Boltzmann constant k	1.380658×10^{-23} (J/K)
Gas constant R	$N_A k = 8.314510$ (J/mol/K)
Standard state condition	0°C, 1 atm

Table B.2 Metric units of pressure (vacuum)

(SI unit) Pa	$\mathbf{N \cdot m^{-2}}$
(Practical and historical units)	
atm (standard atmospheric pressure)	1 atm $\equiv$ 101,325 Pa (definition) $= 1{,}013$ hPa $= 0.10$ MPa
mmHg	height of a mercury column equivalent that exerts the standard atmospheric pressure 1 atm $= 760$ mmHg $= 76.0$ cmHg

(Continued)

Micro- and Nanofabrication for Beginners
Eiichi Kondoh
Copyright © 2021 Jenny Stanford Publishing Pte. Ltd.
ISBN 978-981-4877-09-1 (Hardcover), 978-1-003-11993-7 (eBook)
www.jennystanford.com

Table B.2 (*Continued*)

(SI unit) Pa	$N \cdot m^{-2}$
Torr	760 Torr is defined as 760 mmHg. (introduced to measure low pressures to avert the effects of Hg vapor pressure and surface tension) 1 Torr = 1 mmHg = 133.322 Pa
mmH_2O	height of a water column equivalent that exerts the standard atmospheric pressure 1 atm = 10,332 mmH_2O
bar	1 bar = 10^5 Pa = 10^6 dyne·cm^{-2}

Table B.3 Conversion table of vacuum pressure units

	Pa	atm	kgf/cm^2	bar	Torr (mmHg)	mmH$_2$O
1 atm	101,325	1	1.0332	1.0133	760	103,320
1 kgf/cm^2	98066	0.96784	1	0.9807	735.6	10^5
1 bar	10^5	0.98692	1.0197	1	750.06	101,972
1 Torr (mmHg)	133.32				1	135.95
1 mmH$_2$O	0.98066					1

Appendix C

Development of Facets

Euhedral crystals have a well-formed, clear, and sharply-edged faces (facets). These shapes develop reflecting the arrangements of atoms or the crystal structures (Fig. C.1). The fact that crystals consist of periodically arranged atoms was found from the euhedral shapes in ca 1800s. Figure C.2 is a picture of a giant quartz crystal. It shows

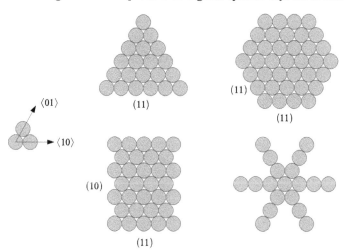

Figure C.1 Atomic structures and facetting.

Micro- and Nanofabrication for Beginners
Eiichi Kondoh
Copyright © 2021 Jenny Stanford Publishing Pte. Ltd.
ISBN 978-981-4877-09-1 (Hardcover), 978-1-003-11993-7 (eBook)
www.jennystanford.com

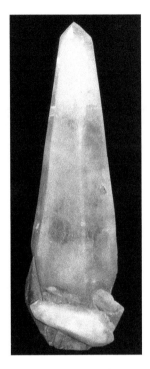

Figure C.2 Giant (97 cm high) eudhedral quartz. Collection of quartz museum of University of Yamanashi.

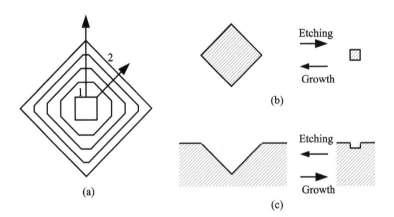

Figure C.3 Development of facets.

a well-developed hexahedral shape, as the quartz has a hexagonal structure.

Let us see in Fig. C.3. that an euhedral shape is determined by the growth or etching rate of crystal planes. Now we assume that the innermost square in Fig. C.3(a) is the initial shape of a crystal. When plane 1 grows faster than plane 2, or the growth rate to direction 1 is larger than that of direction 2, plane 1 disappears gradually and only plane 2 remains. Etching is the opposite of the growth. When the outermost crystal shape is etched, the plane etched fastest remains and appears finally as an euhedral fact. That is, for a concave-shaped crystal, the slowly growing plane or the fast-etched planes will appear [Fig. C.3(c)].

Next, we assume that the diagrams of Fig. C.3(a) show the shapes of voids. When a void grows, or the void volume becomes larger, the slowest etched plane 2 appears, whereas when a void shrinks, the fastest growing plane 1 appears. Therefore, for a convex-shaped crystal, the fast-growing planes or the slowly-etched planes will appear.

Index